T0213485

SpringerBriefs in Mathematics

Series Editors

Nicola Bellomo, Torino, Italy

Michele Benzi, Pisa, Italy

Palle Jorgensen, Iowa, USA

Tatsien Li, Shanghai, China

Roderick Melnik, Waterloo, Canada

Otmar Scherzer, Linz, Austria

Benjamin Steinberg, New York, USA

Lothar Reichel, Kent, USA

Yuri Tschinkel, New York, USA

George Yin, Detroit, USA

Ping Zhang, Kalamazoo, USA

SpringerBriefs present concise summaries of cutting-edge research and practical applications across a wide spectrum of fields. Featuring compact volumes of 50 to 125 pages, the series covers a range of content from professional to academic. Briefs are characterized by fast, global electronic dissemination, standard publishing contracts, standardized manuscript preparation and formatting guidelines, and expedited production schedules.

Typical topics might include:

A timely report of state-of-the art techniques A bridge between new research results, as published in journal articles, and a contextual literature review A snapshot of a hot or emerging topic An in-depth case study A presentation of core concepts that students must understand in order to make independent contributions

SpringerBriefs in Mathematics showcases expositions in all areas of mathematics and applied mathematics. Manuscripts presenting new results or a single new result in a classical field, new field, or an emerging topic, applications, or bridges between new results and already published works, are encouraged. The series is intended for mathematicians and applied mathematicians. All works are peer-reviewed to meet the highest standards of scientific literature.

Titles from this series are indexed by Scopus, Web of Science, Mathematical Reviews, and zbMATH.

More information about this series at http://www.springer.com/series/10030

Atsushi Yagi

Abstract Parabolic Evolution Equations and Łojasiewicz–Simon Inequality I

Abstract Theory

Springer

Atsushi Yagi
Osaka University
Suita, Osaka, Japan

ISSN 2191-8198 ISSN 2191-8201 (electronic)
SpringerBriefs in Mathematics
ISBN 978-981-16-1895-6 ISBN 978-981-16-1896-3 (eBook)
https://doi.org/10.1007/978-981-16-1896-3

© The Author(s), under exclusive license to Springer Nature Singapore Pte Ltd. 2021
This work is subject to copyright. All rights are solely and exclusively licensed by the Publisher, whether
the whole or part of the material is concerned, specifically the rights of translation, reprinting, reuse
of illustrations, recitation, broadcasting, reproduction on microfilms or in any other physical way, and
transmission or information storage and retrieval, electronic adaptation, computer software, or by similar
or dissimilar methodology now known or hereafter developed.
The use of general descriptive names, registered names, trademarks, service marks, etc. in this publication
does not imply, even in the absence of a specific statement, that such names are exempt from the relevant
protective laws and regulations and therefore free for general use.
The publisher, the authors, and the editors are safe to assume that the advice and information in this book
are believed to be true and accurate at the date of publication. Neither the publisher nor the authors or
the editors give a warranty, expressed or implied, with respect to the material contained herein or for any
errors or omissions that may have been made. The publisher remains neutral with regard to jurisdictional
claims in published maps and institutional affiliations.

This Springer imprint is published by the registered company Springer Nature Singapore Pte Ltd.
The registered company address is: 152 Beach Road, #21-01/04 Gateway East, Singapore 189721,
Singapore

Preface

After publishing his monograph [Yag10], the present author together with his collaborators studied convergence of global solutions to equilibria as time goes to infinity for some concrete parabolic equations, such as three-dimensional Laplace reaction–diffusion equations, epitaxial growth equations incorporating logarithmic roughening functions, and one-dimensional Keller–Segel equations, by making use of the Łojasiewicz–Simon gradient inequality. Through these studies, he found out that by using the techniques of abstract parabolic evolution equations explained in his monograph and the functional analytic approach to the gradient inequality due to Chill [Chi06] and Haraux–Jendoubi [HJ11], one can set up some general frameworks and treat these equations in a unified way. This monograph and the subsequent monograph are then addressed to explaining the general results in an abstract style and to showing how to apply the results to various model equations from the real world.

We consider a nonlinear abstract evolution equation

$$u' + A(u) = 0, \qquad 0 < t < \infty, \tag{EE}$$

in an infinite-dimensional space. As the prerequisite assumptions, we assume that there exists a global solution $u(t)$ to (EE) and that there is a Lyapunov function $\Phi: X \to \mathbb{R}$ for which it holds that

$$\frac{d}{dt} \Phi(u(t)) < 0 \qquad \text{for any } 0 \leq t < \infty. \tag{i}$$

More precisely, let Z and X be two real Hilbert spaces such that $Z \subset X$ densely and compactly and let $Z \subset X \subset Z^*$ be a triplet of spaces, Z^* being an adjoint space of Z. Then, we assume that A is a continuous nonlinear operator from Z into Z^*, that the solution $u(t)$ belongs to $\mathscr{C}([0, \infty); Z) \cap \mathscr{C}^1([0, \infty); Z^*)$ together with the global boundedness

$$\sup_{0 \leq t < \infty} \|u(t)\|_Z < \infty, \tag{ii}$$

and that $\Phi(u)$ is differentiable on X with derivative $\dot{\Phi}(u)$ in the sense that

$$\Phi(u+h) - \Phi(u) = (\dot{\Phi}(u), h)_X + o(\|h\|_X) \qquad \text{as } \|h\|_X \to 0,$$

$\dot{\Phi}(u)$ being continuous from both X and Z into themselves, respectively. The motivation for such settings is described in the beginning of Chap. 2.

It is, however, known that there is a counter example showing that, even if $\Phi : X \to \mathbb{R}$ is smooth, (i) and (ii) may not imply that $u(t)$ is convergent in Z^* as $t \to \infty$. In other words, (i) and (ii) may not imply that the ω-limit set

$$\omega(u) = \{\overline{u} \in Z^*; \ \exists t_n \nearrow \infty \text{ such that } u(t_n) \to \overline{u} \text{ in } Z^*\}$$

of $u(t)$ is a singleton. As a general condition which can provide, together with (i) and (ii), the desired convergence of $u(t)$, the following condition

$$\|\dot{\Phi}(u(t))\|_Z \geq D|\Phi(u(t)) - \Phi(\overline{u})|^{1-\theta} \qquad \text{if } u(t) \in B^X(\overline{u}; r), \tag{iii}$$

is known, where $\overline{u} \in \omega(u)$ denotes an ω-limit satisfying $\dot{\Phi}(\overline{u}) = 0$ and θ denotes an exponent such that $0 < \theta \leq \frac{1}{2}$.

For a $\mathscr{C}^1$ function $\Phi : X \to \mathbb{R}$ and its critical point $\overline{u}$ (i.e., $\dot{\Phi}(\overline{u}) = 0$), when there exists an exponent $0 < \theta \leq \frac{1}{2}$ for which the estimate

$$\|\dot{\Phi}(u)\|_X \geq D|\Phi(u) - \Phi(\overline{u})|^{1-\theta} \qquad \text{if } u \in B^X(\overline{u}; r), \tag{ŁS}$$

holds true in a neighborhood of $\overline{u}$, $D > 0$ being a constant and $r > 0$ being a radius, this estimate is called the Łojasiewicz–Simon gradient inequality. The inequality (ŁS) was first verified in 1962 by Łojasiewicz [Loj63] in the case where $\dim X < \infty$ under the assumption that $\Phi(u)$ is analytic in a neighborhood of $\overline{u}$. Twenty years later, Simon devised methods to verify (ŁS) in the case where $\dim X = \infty$ through some concrete partial differential equations and corresponding Lyapunov functions. Afterwards, Simon's methods were developed by many researchers. Among others, Chill [Chi06] and Haraux–Jendoubi [HJ11] presented the general results providing (ŁS) for $\Phi(u)$ in Hilbert spaces apart from concrete partial differential equations, and at the same time they showed ways to apply their results to partial differential equations.

In order to invent methods for verifying (iii) for the solutions of (EE), which are available to various concrete equations, one can basically follow the methods due to Chill and Haraux–Jendoubi. But, we have to clear up some technical assumptions on the solutions $u(t)$ and on the Lyapunov functions $\Phi(u)$ which are needed in their arguments. For this reason, first, we shall somewhat modify their settings for obtaining (iii) by introducing another suitable space Y, which plays an important role in addition to the underlying spaces $Z \subset X \subset Z^*$. Second, we shall clear up those technical assumptions by utilizing the techniques devised in [Yag10].

Not only shall we clear up the technical assumptions, but also our methods in fact will provide more precise and wider application results. As will be shown in [Yag], we can prove the convergence for global solutions of semilinear parabolic equations assuming only that they are globally bounded with respect to the spatial L_∞-norm under some minimum regularity conditions on the spatial domains, can handle epitaxial growth models with general roughening functions, and can handle higher dimensional Keller–Segel equations which are known as typical model equations in chemotaxis. The author believes that the methods explained here are applicable to other various partial differentiable equations of parabolic type. Some comments for applications are made at the end of Chaps. 2 and 3.

There are some other methods for developing the theory of the Łojasiewicz–Simon gradient inequality, see Feireisl–Issard–Roch–Petzeltová [FIP04], Huang [Hua06], and so on. But those results often provide no precise order estimate of convergence of $u(t)$ to $\overline{u}$. By the refined arguments, our method can give an order estimate of convergence of the form

$$\|u(t) - \overline{u}\|_{Z^*} \leq C[\Phi(u(t)) - \Phi(\overline{u})]^\theta, \qquad \text{as } t \to \infty, \qquad \text{(iv)}$$

C being some constant. Since it is not so difficult in applications to know a profile of the monotonously decreasing function $\Phi(u(t))$ for $0 < t < \infty$, (iv) provides us some information about convergence of $u(t)$ to the stationary solution $\overline{u}$.

In order to keep the brief monographs in a suitable length, the contents are divided into two volumes. This monograph (Volume 1) mainly addresses explaining the abstract theory. Chapter 1 prepares all basic matters that are used in the monograph. Especially, the notions of real subspaces and real sectorial operators are new and very convenient for utilizing the techniques devised in complex functional analysis in the real frameworks. Chapter 2 sets up the frameworks of our study and makes four structural assumptions including (iii), which is of course the most crucial one. The asymptotic convergence (iv) of the global solutions of (EE) to stationary solutions as time goes to infinity is then derived immediately under the four structural assumptions. Chapter 3 is devoted to inventing new methods for verifying (iii) on the basis of those due to [Chi06] and [HJ11].

Applications of these abstract results to the concrete partial differential equations mentioned above are all presented in the subsequent monograph [Yag].

To read this monograph, knowledge of functional analysis of standard level is required ([Bre11, Yos80, Zei88] and so on). The readers are also expected to be familiar with the functional analytic methods for partial differential equations ([DL84a, DL84b, Tan75, Tan97, Yag10] and so on).

Suita, Osaka, Japan   Atsushi Yagi
December 2020

Contents

Chapter 1
Preliminaries

In this monograph, we mainly handle real Banach spaces and Hilbert spaces and real linear operators. For this reason, Banach spaces, Hilbert spaces and linear operators always mean real ones if they are not otherwise specified. However, they are all obtained as a real part (in a certain sense) of some complex Banach spaces, Hilbert spaces or complex linear operators. This means that we can fortunately utilize the powerful tools developed in Complex Functional Analysis. The first half of this chapter is devoted to reviewing the most convenient way to utilize the tools of Complex Functional Analysis to real spaces and operators. These results were originally presented in the paper [Yag17].

After reviewing the basic materials of Complex Functional Analysis in Sect. 1.1, a notion of real subspaces of complex Banach spaces is introduced in Sect. 1.2 and a notion of real operators in complex Banach spaces is introduced in Sect. 1.3. Utilizing the notions, we give a definition of realizations of elliptic operators in real function spaces and show their basic properties in Sect. 1.4.

The second half of this chapter is devoted to reviewing some results concerning Fréchet and Gâteaux differentiability for nonlinear operators from a Banach space into another.

This chapter concludes by recalling the definition of Fredholm operators and their important class which is given by Riesz–Schauder theory and by recalling the Łojasiewicz gradient inequality for analytic functions.

1.1 Basic Materials in Complex Functional Analysis

All the results in this section were already explained or obtained in [Yag10].

Let $X_{\mathbb{C}}$ and $Y_{\mathbb{C}}$ be two Banach spaces with norm $\| \cdot \|_X$ and $\| \cdot \|_Y$ respectively over the complex number $\mathbb{C}$.

© The Author(s), under exclusive license to Springer Nature Singapore Pte Ltd. 2021
A. Yagi, *Abstract Parabolic Evolution Equations and Łojasiewicz-Simon Inequality I*,
SpringerBriefs in Mathematics, https://doi.org/10.1007/978-981-16-1896-3_1

Let $\mathscr{L}(X_{\mathbb{C}}, Y_{\mathbb{C}})$ be the space of all bounded linear operators from $X_{\mathbb{C}}$ into $Y_{\mathbb{C}}$. For each $T \in \mathscr{L}(X_{\mathbb{C}}, Y_{\mathbb{C}})$, its uniform operator norm is defined by $\|T\|_{\mathscr{L}(X,Y)} = \sup_{\|f\|_X \leq 1} \|Tf\|_Y$. Then, $\mathscr{L}(X_{\mathbb{C}}, Y_{\mathbb{C}})$ becomes a complex Banach space. When $X_{\mathbb{C}} = Y_{\mathbb{C}}$, $\mathscr{L}(X_{\mathbb{C}}, X_{\mathbb{C}})$ is written as $\mathscr{L}(X_{\mathbb{C}})$ for short. When $Y_{\mathbb{C}} = \mathbb{C}$, the bounded operator $\varphi \in \mathscr{L}(X_{\mathbb{C}}, \mathbb{C})$ is called a bounded linear functional on $X_{\mathbb{C}}$.

1.1.1 Dual Spaces

In $\mathscr{L}(X_{\mathbb{C}}, \mathbb{C})$, we prefer to define a multiplication between the complex number α and the bounded linear functional φ as

$$[\alpha\varphi](f) = \overline{\alpha}\varphi(f) \qquad \text{for } f \in X_{\mathbb{C}}, \tag{1.1}$$

rather than to define $[\alpha\varphi](f) = \alpha\varphi(f)$, $f \in X_{\mathbb{C}}$, as usual. When $\mathscr{L}(X_{\mathbb{C}}, \mathbb{C})$ is equipped with this multiplication between the complex numbers and the bounded linear functionals, the space is denoted by $X'_{\mathbb{C}}$ which is a complex linear space. Furthermore, by the uniform operator norm mentioned above, $X'_{\mathbb{C}}$ also becomes a Banach space which we call *the dual space* of $X_{\mathbb{C}}$.

The definition (1.1) sometimes provides us an advantage over the usual one. Let $X_{\mathbb{C}}$ be a complex Hilbert space with inner product $(\cdot, \cdot)_X$. Then, for each $g \in X_{\mathbb{C}}$, the linear functional $(\cdot, g)_X$ is continuous on $X_{\mathbb{C}}$; therefore, a correspondence $J : g \mapsto Jg$ from $X_{\mathbb{C}}$ into $\mathscr{L}(X_{\mathbb{C}}, \mathbb{C})$ is defined by the relation

$$[Jg](f) = (f, g)_X \qquad \text{for all } f \in X_{\mathbb{C}}. \tag{1.2}$$

By virtue of the Riesz representation theorem, $J : X_{\mathbb{C}} \to \mathscr{L}(X_{\mathbb{C}}, \mathbb{C})$ is a one-to-one and onto mapping and an isometry, but J is conjugate linear, i.e., $J(\alpha g) = \overline{\alpha}Jg$. However, we verify that J is *linear* from $X_{\mathbb{C}}$ onto $X'_{\mathbb{C}}$ owing to (1.1).

Theorem 1.1 *If $X_{\mathbb{C}}$ is a Hilbert space, then the mapping $J : X_{\mathbb{C}} \to X'_{\mathbb{C}}$ defined by* (1.2) *is an isometric isomorphism.*

1.1.2 Adjoint Spaces

Let $X_{\mathbb{C}}$ and $Y_{\mathbb{C}}$ be complex Banach spaces with norms $\|\cdot\|_X$ and $\|\cdot\|_Y$ respectively. A complex-valued function $\langle \cdot, \cdot \rangle$ defined on the product space $X \times Y$ having the properties

$$\begin{cases} \langle \alpha f + \beta \tilde{f}, g \rangle = \alpha \langle f, g \rangle + \beta \langle \tilde{f}, g \rangle, & \alpha, \beta \in \mathbb{C}, \ f, \tilde{f} \in X_{\mathbb{C}}, \ g \in Y_{\mathbb{C}}, \\ \langle f, \alpha g + \beta \tilde{g} \rangle = \overline{\alpha} \langle f, g \rangle + \overline{\beta} \langle f, \tilde{g} \rangle, & \alpha, \beta \in \mathbb{C}, \ f \in X_{\mathbb{C}}, \ g, \tilde{g} \in Y_{\mathbb{C}}, \end{cases} \tag{1.3}$$

is called *a sesquilinear form* on $X_\mathbb{C} \times Y_\mathbb{C}$. Furthermore, a sesquilinear form on $X_\mathbb{C} \times Y_\mathbb{C}$ is called *a duality product* if it satisfies

$$|\langle f, g \rangle| \le \|f\|_X \|g\|_Y, \qquad f \in X_\mathbb{C}, \ g \in Y_\mathbb{C},$$

$$\|f\|_X = \sup_{\|g\|_Y \le 1} |\langle f, g \rangle|, \qquad f \in X_\mathbb{C},$$

$$\|g\|_Y = \sup_{\|f\|_X \le 1} |\langle f, g \rangle|, \qquad g \in Y_\mathbb{C}.$$

When there exists such a duality product for two Banach spaces $X_\mathbb{C}$ and $Y_\mathbb{C}$, the space $Y_\mathbb{C}$ is called *an adjoint space* of $X_\mathbb{C}$. Obviously, this relation is symmetric, that is, if $Y_\mathbb{C}$ is adjoint to $X_\mathbb{C}$ with $\langle \cdot, \cdot \rangle_{X \times Y}$, then $X_\mathbb{C}$ is adjoint to $Y_\mathbb{C}$ with the duality product $\langle g, f \rangle_{Y \times X} = \overline{\langle f, g \rangle}_{X \times Y}$. This means that adjointness is a property for pairs of Banach spaces and it is allowed to say that $\{X_\mathbb{C}, Y_\mathbb{C}\}$ forms *an adjoint pair* with duality product $\langle \cdot, \cdot \rangle$.

If X is a Hilbert space, then $X_\mathbb{C}$ is clearly adjoint to itself with its inner product $(\cdot, \cdot)_X$. Hence, $\{X_\mathbb{C}, X_\mathbb{C}\}$ is an adjoint pair.

Let $X_\mathbb{C}$ be a Banach space, and $X'_\mathbb{C}$ be its dual space. We introduce the function $\langle f, \varphi \rangle = \varphi(f)$ for $f \in X_\mathbb{C}$ and $\varphi \in X'_\mathbb{C}$. Owing to (1.1), this function defines a sesquilinear form on $X_\mathbb{C} \times X'_\mathbb{C}$. Moreover, by virtue of the Hahn–Banach extension theorem, this function gives a duality product on $X_\mathbb{C} \times X'_\mathbb{C}$. Hence, $\{X_\mathbb{C}, X'_\mathbb{C}\}$ forms an adjoint pair.

As verified above, $\{X_\mathbb{C}, X'_\mathbb{C}\}$ is an adjoint pair. So is $\{X'_\mathbb{C}, X''_\mathbb{C}\}$. Therefore, $X'_\mathbb{C}$ has at least two adjoint spaces $X_\mathbb{C}$ and $X''_\mathbb{C}$ if $X_\mathbb{C}$ is a non-reflexive space. In order to denote an unspecified adjoint space of $X_\mathbb{C}$ we shall use the notation $X^*_\mathbb{C}$. Therefore, $X^*_\mathbb{C}$ is not uniquely determined from $X_\mathbb{C}$, quite differently from its dual space $X'_\mathbb{C}$.

Let $\{X_\mathbb{C}, X^*_\mathbb{C}\}$ be an adjoint pair with duality product $\langle \cdot, \cdot \rangle_{X \times X^*}$. For each $g \in X^*_\mathbb{C}$, the linear functional $\langle \cdot, g \rangle_{X \times X^*}$ is continuous on $X_\mathbb{C}$; therefore, a correspondence $J : g \mapsto Jg$ from $X^*_\mathbb{C}$ into $X'_\mathbb{C}$ is defined by the relation

$$[Jg](f) = \langle f, g \rangle_{X \times X^*} \qquad \text{for all } f \in X_\mathbb{C}. \tag{1.4}$$

Owing to (1.1), J is a linear operator on $X^*_\mathbb{C}$. In addition, it is easy to verify that J is an isometry from $X^*_\mathbb{C}$ onto $J(X^*_\mathbb{C}) \subset X'_\mathbb{C}$. Furthermore, when $X_\mathbb{C}$ is a reflexive Banach space, J is verified to be onto $X'_\mathbb{C}$.

Theorem 1.2 *If $X_\mathbb{C}$ is a reflexive Banach space, then the mapping $J : X^*_\mathbb{C} \to X'_\mathbb{C}$ defined by (1.4) is an isometric isomorphism.*

When $X_\mathbb{C}$ is a Hilbert space, $\{X_\mathbb{C}, X_\mathbb{C}\}$ is an adjoint pair with its inner product. So, the two definitions (1.2) and (1.4) for the operator J are consistent.

1.1.3 Interpolation of Spaces

Let $X_{\mathbb{C}}$ and $Z_{\mathbb{C}}$ be two complex Banach spaces with norm $\|\cdot\|_X$ and $\|\cdot\|_Z$ respectively. Assume that $Z_{\mathbb{C}} \subset X_{\mathbb{C}}$ densely and continuously.

Let $S = \{z; \ 0 < \operatorname{Re} z < 1\}$ be the strip in the plane $\mathbb{C}$. By $\mathscr{H}(X_{\mathbb{C}}, Z_{\mathbb{C}})$ we denote a space of analytic functions with the following properties:

$$\mathscr{H}(X_{\mathbb{C}}, Z_{\mathbb{C}}) = \{F(z); \ F(z) \text{ is an analytic function for } z \in S \text{ with values in } X_{\mathbb{C}},$$

$$\text{is a bounded and continuous function for } z \in \overline{S} \text{ with values in } X_{\mathbb{C}},$$

$$\text{and is a bounded and continuous function for } z = 1 + iy \text{ with}$$

$$\text{values in } Z_{\mathbb{C}}\}.$$

For $F \in \mathscr{H}(X_{\mathbb{C}}, Z_{\mathbb{C}})$, its norm is defined by

$$\|F\|_{\mathscr{H}} = \max \left\{ \sup_{-\infty < y < \infty} \|F(iy)\|_X, \ \sup_{-\infty < y < \infty} \|F(1+iy)\|_Z \right\}.$$

Then, $\mathscr{H}(X_{\mathbb{C}}, Z_{\mathbb{C}})$ is a Banach space.

For $0 \le \theta \le 1$, we define the space $[X_{\mathbb{C}}, Z_{\mathbb{C}}]_\theta$ in the following way:

$$[X_{\mathbb{C}}, Z_{\mathbb{C}}]_\theta = \{u \in X_{\mathbb{C}}; \ \text{there exists a function } F \in \mathscr{H}(X_{\mathbb{C}}, Z_{\mathbb{C}})$$

$$\text{such that } u = F(\theta)\}.$$

For $u \in [X_{\mathbb{C}}, Z_{\mathbb{C}}]_\theta$, its norm is defined by

$$\|u\|_\theta = \inf_{F \in \mathscr{H}, \ F(\theta)=u} \|F\|_{\mathscr{H}}. \tag{1.5}$$

Then, $[X_{\mathbb{C}}, Z_{\mathbb{C}}]_\theta$ is a Banach space with the norm $\|\cdot\|_\theta$ and possesses the following basic properties:

1. $[X_{\mathbb{C}}, Z_{\mathbb{C}}]_0 = X_{\mathbb{C}}$ and $[X_{\mathbb{C}}, Z_{\mathbb{C}}]_1 = Z_{\mathbb{C}}$ with isometries.
2. For $0 < \theta < 1$, $Z_{\mathbb{C}} \subset [X_{\mathbb{C}}, Z_{\mathbb{C}}]_\theta \subset X_{\mathbb{C}}$ densely and continuously.
3. For $0 < \theta < 1$, $\|u\|_\theta \le \|u\|_X^{1-\theta} \|u\|_Z^{\theta}$ for all $u \in Z_{\mathbb{C}}$.
4. For $0 < \theta < \theta' < 1$, $[X_{\mathbb{C}}, Z_{\mathbb{C}}]_{\theta'} \subset [X_{\mathbb{C}}, Z_{\mathbb{C}}]_\theta$ continuously.

1.1.4 Triplets of Spaces

Let $X_{\mathbb{C}}$ be a complex Hilbert space with inner product $(\cdot, \cdot)_X$ and norm $\|\cdot\|_X$ and let $Z_{\mathbb{C}}$ be a second complex Hilbert space with norm $\|\cdot\|_Z$. Assume that $Z_{\mathbb{C}} \subset X_{\mathbb{C}}$ densely and continuously.

By the techniques of extrapolation of spaces, we can construct a third space $Z_{\mathbb{C}}^*$ which is a complex Banach space with norm $\| \cdot \|_{Z^*}$ and enjoys the following properties:

1. $Z_{\mathbb{C}} \subset X_{\mathbb{C}} \subset Z_{\mathbb{C}}^*$ with dense and continuous embeddings.
2. $\{Z_{\mathbb{C}}, Z_{\mathbb{C}}^*\}$ forms an adjoint pair with duality product $\langle \cdot, \cdot \rangle_{Z \times Z^*}$, i.e.,

$$|\langle u, \varphi \rangle_{Z \times Z^*}| \leq \|u\|_Z \|\varphi\|_{Z^*}, \qquad u \in Z_{\mathbb{C}}, \ \varphi \in Z_{\mathbb{C}}^*, \tag{1.6}$$

$$\|u\|_Z = \sup_{\|\varphi\|_{Z^*} \leq 1} |\langle u, \varphi \rangle_{Z \times Z^*}|, \qquad u \in Z_{\mathbb{C}}, \tag{1.7}$$

$$\|\varphi\|_{Z^*} = \sup_{\|u\|_Z \leq 1} |\langle u, \varphi \rangle_{Z \times Z^*}|, \qquad \varphi \in Z_{\mathbb{C}}^*. \tag{1.8}$$

3. The duality product satisfies

$$\langle u, f \rangle_{Z \times Z^*} = (u, f)_X, \qquad u \in Z_{\mathbb{C}}, \ f \in X_{\mathbb{C}}. \tag{1.9}$$

Such a space $Z_{\mathbb{C}}^*$ is, as a matter of fact, uniquely determined from $Z_{\mathbb{C}} \subset X_{\mathbb{C}}$. We call the spaces $Z_{\mathbb{C}} \subset X_{\mathbb{C}} \subset Z_{\mathbb{C}}^*$ a *triplet of complex spaces*.

We also notice some important properties of triplets. The property (1) naturally yields inclusions for their dual spaces such that $(Z_{\mathbb{C}}^*)' \subset X_{\mathbb{C}}' \subset Z_{\mathbb{C}}'$ densely and continuously. Let J be the isometric isomorphism defined by (1.2) or (1.4). Then, J is an isometric isomorphism not only from $X_{\mathbb{C}}$ onto $X_{\mathbb{C}}'$ and from $Z_{\mathbb{C}}^*$ onto $Z_{\mathbb{C}}'$ but also from $Z_{\mathbb{C}}$ onto $(Z_{\mathbb{C}}^*)'$ and it holds that

$$\begin{cases} [J\varphi](u) = \langle u, \varphi \rangle_{Z \times Z^*} & \text{for any} \quad u \in Z_{\mathbb{C}}, \ \varphi \in Z_{\mathbb{C}}^*, \\ [Jg](f) = (f, g)_X & \text{for any} \quad f \in X_{\mathbb{C}}, \ g \in X_{\mathbb{C}}, \\ [Ju](\varphi) = \langle \varphi, u \rangle_{Z^* \times Z} & \text{for any} \quad u \in Z_{\mathbb{C}}, \ \varphi \in Z_{\mathbb{C}}^*. \end{cases}$$

The space $X_{\mathbb{C}}$ is always characterized as the interpolation space

$$[Z_{\mathbb{C}}^*, Z_{\mathbb{C}}]_{\frac{1}{2}} = X_{\mathbb{C}} \qquad \text{with norm equivalence.} \tag{1.10}$$

Remark 1.1 As for constructing the triplet, it may be more straightforward to use the identification of $X_{\mathbb{C}}$ and its dual $X_{\mathbb{C}}'$. The dense and continuous embedding $Z_{\mathbb{C}} \subset X_{\mathbb{C}}$ naturally implies that $X_{\mathbb{C}}' \subset Z_{\mathbb{C}}'$ densely and continuously. So, the identification yields a triplet $Z_{\mathbb{C}} \subset X_{\mathbb{C}} \approx X_{\mathbb{C}}' \subset Z_{\mathbb{C}}'$. However, here we used another method. Specifying the desired properties of the triplet in advance, we constructed step by step the third space $Z_{\mathbb{C}}^*$ by the techniques of extrapolation. As mentioned, the space $Z_{\mathbb{C}}^*$ having the properties (1)–(3) is uniquely determined. So, both $Z_{\mathbb{C}} \subset X_{\mathbb{C}} \approx X_{\mathbb{C}}' \subset Z_{\mathbb{C}}'$ and $Z_{\mathbb{C}} \subset X_{\mathbb{C}} \subset Z_{\mathbb{C}}^*$ provide us an equivalent structure of three spaces.

1.1.5   Sectorial Operators

Let $X_\mathbb{C}$ be a complex Banach space with norm $\|\cdot\|_X$. A densely defined, closed linear operator A of $X_\mathbb{C}$ is called *a sectorial operator* if its spectrum $\sigma(A)$ is contained in an open sectorial domain

$$\sigma(A) \subset \Sigma_\omega = \{\lambda \in \mathbb{C}; \ |\arg\lambda| < \omega\}, \qquad 0 < \omega \le \pi, \tag{1.11}$$

and if its resolvent satisfies the estimate

$$\|(\lambda - A)^{-1}\| \le \frac{M}{|\lambda|} \qquad \text{for } \lambda \notin \Sigma_\omega, \tag{1.12}$$

with some constant $M \ge 1$.

Since (1.11) implicitly means that $0 \notin \sigma(A)$, i.e., $0 \in \rho(A)$, the condition implies that A has a bounded inverse $A^{-1} \in \mathscr{L}(X_\mathbb{C})$. Under (1.11)–(1.12), if $\omega'(< \omega)$ is sufficiently close to ω, then there exists a constant $M' > M$ such that (1.11)–(1.12), but ω' and M' substituting for ω and M, are valid. So, the infimum of the angles ω for which (1.11)–(1.12) hold true is called the angle of A and is denoted by ω_A. Then, we have $0 \le \omega_A < \pi$ and $\sigma(A) \subset \overline{\Sigma}_{\omega_A}$.

(I) Fractional Powers Let A be a sectorial operator of $X_\mathbb{C}$ and let (1.11)–(1.12) be satisfied with $0 < \omega < \pi$. For $\mathrm{Re}\, z > 0$, its fractional power of exponent $-z$ is defined by the integral

$$A^{-z} = \frac{1}{2\pi i} \int_\Gamma \lambda^{-z} (\lambda - A)^{-1} d\lambda \tag{1.13}$$

in the space $\mathscr{L}(X_\mathbb{C})$, $\Gamma = \Gamma_- \cup \Gamma_\delta \cup \Gamma_+$ being an integral contour lying in $\rho(A)$ given by $\Gamma_\pm : \lambda = re^{\pm i\omega}$, $\delta \le r < \infty$ and $\Gamma_\delta : \lambda = \delta e^{i\theta}$, $-\omega \le \theta \le \omega$, where $\delta > 0$ is a sufficiently small number such that $\{\lambda; \ |\lambda| \le \delta\} \subset \rho(A)$.

Analyticity of A^{-z} for $\mathrm{Re}\, z > 0$ and invertibility of A^{-1} implies that the bounded linear operator A^{-z} is also invertible for every $\mathrm{Re}\, z > 0$; therefore, $A^z = [A^{-z}]^{-1}$ is defined as a closed linear operator of $X_\mathbb{C}$.

We can list the basic properties of the fractional powers A^x $(-\infty < x < \infty)$:

1. For $0 < x < \infty$, A^{-x} is an analytic function with values in $\mathscr{L}(X_\mathbb{C})$.
2. It holds that $A^{-x} A^{-x'} = A^{-(x+x')}$ for any $0 < x, x' < \infty$.
3. For $0 < x < \infty$, A^x is a densely defined, closed linear operator of $X_\mathbb{C}$.
4. It holds that $A^x A^{x'} = A^{x+x'}$ for any $0 < x, x' < \infty$ in the sense of product for unbounded linear operators.
5. For each $f \in X_\mathbb{C}$, as $x \searrow 0$, $A^{-x} f$ tends to f in $X_\mathbb{C}$.

In view of the property (5), we are led to define $A^0 = I$.

Let $0 < \theta < 1$. It is possible to show that A^{θ} is also a sectorial operator of X with angle $\omega_{A^{\theta}} \leq \theta \omega_A$. So, the fractional powers of A^{θ} are considered; indeed, we can verify that

$$[A^{\theta}]^{-x} = A^{-\theta x} \quad \text{and} \quad [A^{\theta}]^x = A^{\theta x} \quad \text{for} \quad 0 < x < \infty. \tag{1.14}$$

For $0 < \theta < 1$, it is known that the inequality $\|A^{\theta} u\|_X \leq C \|u\|_X^{1-\theta} \|Au\|_X^{\theta}$ holds for $u \in \mathscr{D}(A)$. In view of this and the inequality (3) of interpolation, one may expect that $\mathscr{D}(A^{\theta})$ would be characterized by $[X_{\mathbb{C}}, \mathscr{D}(A)]_{\theta}$, here the domain $\mathscr{D}(A)$ is considered as a Banach space equipped with the graph norm $\|u\|_{\mathscr{D}(A)} = \|Au\|_X$. Unfortunately, it is already known that this is not the case in general even if $X_{\mathbb{C}}$ is a Hilbert space. However, some classes of sectorial operators are known for which the interpolation characterization is the case. In fact, when $X_{\mathbb{C}}$ is a Hilbert space, the possibility of continuous extension of function A^{-z} from $\{z; \operatorname{Re} z > 0\}$ into $\mathscr{L}(X_{\mathbb{C}})$ up to the imaginary axis $i\mathbb{R}$ (like positive definite self-adjoint operators) is equivalent to $\mathscr{D}(A^{\theta}) = [X_{\mathbb{C}}, \mathscr{D}(A)]_{\theta}$ and $\mathscr{D}(A^{*\theta}) = [X_{\mathbb{C}}, \mathscr{D}(A^*)]_{\theta}$ for any $0 < \theta < 1$ with norm equivalence. When $X_{\mathbb{C}}$ is a Banach space, the existence of a bounded H_{∞}-functional calculus of A in Σ_{ω} is a sufficient condition for the characterization.

Finally, we want to notice some specific property which we need in [Yag, Chapter 5].

Theorem 1.3 *Let A be a positive definite self-adjoint operator of a Hilbert space $X_{\mathbb{C}}$. Let $0 \leq \theta_0 < \theta_1 \leq 1$. Then, for $\theta_0 < \theta < \theta_1$, it holds that*

$$\mathscr{D}(A^{\theta}) = [\mathscr{D}(A^{\theta_0}), \mathscr{D}(A^{\theta_1})]_{\theta'}, \quad \text{where } (1-\theta')\theta_0 + \theta'\theta_1 = \theta. \tag{1.15}$$

Proof This property, called reiteration, has already been proved by [LM68, Théorème I.6.1]. But we will here give a direct proof by utilizing the exponential law (1.14).

Let $X_i = \mathscr{D}(A^{\theta_i})$ be a Hilbert space equipped with the graph norm $\| \cdot \|_i = \|A^{\theta_i} \cdot \|$ for $i = 0, 1$. Denote by B the linear operator $A^{\theta_1 - \theta_0}$ acting in the space X_0. As $A^{\theta_1 - \theta_0}$ is an isomorphism from X_1 onto X_0, B can be seen as a self-adjoint operator of X_0 with the domain $\mathscr{D}(B) = \mathscr{D}(A^{\theta_1})$. Consequently, it holds for B that $\mathscr{D}(B^{\theta'}) = [X_0, X_1]_{\theta'}$ for any $0 < \theta' < 1$.

On the other hand, B^{θ} can be regarded as a part of $A^{\theta_1 - \theta_0}$ in X_0 (remember that $A^{\theta_1 - \theta_0}$ is originally a self-adjoint operator of X). As a consequence, $B^{\theta'}$ is a part of $[A^{\theta_1 - \theta_0}]^{\theta'}$ in X_0 for any $0 < \theta' < 1$. In particular, $u \in \mathscr{D}(B^{\theta'})$ if and only if $[A^{\theta_1 - \theta_0}]^{\theta'} u \in X_0$, i.e., $u \in \mathscr{D}(A^{\theta_0}[A^{\theta_1 - \theta_0}]^{\theta'})$. We also verify the norm equivalence $\|B^{\theta'} u\|_0 = \|A^{\theta_0}[A^{\theta_1 - \theta_0}]^{\theta'} u\|_X$ for $u \in \mathscr{D}(B^{\theta'})$. In addition, (1.14) yields that $[A^{\theta_1 - \theta_0}]^{\theta'} = A^{(\theta_1 - \theta_0)\theta'}$, i.e., $B^{\theta'}$ is a part of $A^{(\theta_1 - \theta_0)\theta'}$ in X_0. It therefore follows that $\mathscr{D}(B^{\theta'}) = \mathscr{D}(A^{\theta_0} A^{(\theta_1 - \theta_0)\theta'}) = \mathscr{D}(A^{(1-\theta')\theta_0 + \theta'\theta_1})$, together with $\|B^{\theta'} u\|_0 = \|A^{(1-\theta')\theta_0 + \theta'\theta_1} u\|_X$, $u \in \mathscr{D}(B^{\theta'})$. Hence, (1.15) has been proved. $\square$

(II) Analytic Semigroups Let A be a sectorial operator of $X_{\mathbb{C}}$ whose angle is such that $\omega_A < \frac{\pi}{2}$. Then, the analytic semigroup generated by $-A$ is defined by the integral

$$e^{-tA} = \frac{1}{2\pi i} \int_{\Gamma} e^{-t\lambda} (\lambda - A)^{-1} d\lambda, \qquad 0 < t < \infty, \tag{1.16}$$

in the space $\mathscr{L}(X_{\mathbb{C}})$ with an integral contour $\Gamma : \lambda = re^{\pm i\omega}$, $0 \leq r < \infty$, where $\omega_A < \omega < \frac{\pi}{2}$.

The semigroup e^{-tA} has the following basic properties:

1. For $0 < t < \infty$, e^{-tA} is an analytic function with values in $\mathscr{L}(X_{\mathbb{C}})$.
2. It holds that $e^{-tA} e^{-t'A} = e^{-(t+t')A}$ for any $0 < t, t' < \infty$.
3. For $0 < t < \infty$, $\mathscr{R}(e^{-tA}) \subset \mathscr{D}(A)$ and the operator $Ae^{-tA} = e^{-tA}A$ is bounded on $X_{\mathbb{C}}$.
4. The norm estimates $\|e^{-tA}\|_{\mathscr{L}(X)} \leq Ce^{-\delta t}$ and $\|Ae^{-tA}\|_{\mathscr{L}(X)} \leq C(1+t^{-1})e^{-\delta t}$ hold for $0 < t < \infty$ with some $\delta > 0$ and some $C \geq 1$.
5. It holds that $\frac{d}{dt} e^{-tA} = -Ae^{-tA}$ for $0 < t < \infty$.
6. For each $f \in X_{\mathbb{C}}$, as $t \searrow 0$, $e^{-tA} f$ tends to f in $X_{\mathbb{C}}$.

In view of the property (6), we naturally define that $e^{-0A} = I$.

1.1.6 Abstract Parabolic Evolution Equations

Let $X_{\mathbb{C}}$ be a complex Banach space with norm $\|\cdot\|_X$. Let A be a sectorial operator of $X_{\mathbb{C}}$ whose angle is such that $\omega_A < \frac{\pi}{2}$. Consider the Cauchy problem for a semilinear evolution equation

$$\begin{cases} u' + Au = f(u), & 0 < t < \infty, \\ u(0) = u_0, \end{cases} \tag{1.17}$$

in the space $X_{\mathbb{C}}$. Here, $f : \mathscr{D}(f) \to X_{\mathbb{C}}$ is a given nonlinear operator, u_0 is an initial value in $X_{\mathbb{C}}$, and $u = u(t)$ is an unknown function with values in $X_{\mathbb{C}}$.

For the operator f, assume that $\mathscr{D}(A^\eta) \subset \mathscr{D}(f)$ with some exponent $0 \leq \eta < 1$ and assume that f satisfies the following Lipschitz condition:

$$\|f(u) - f(v)\|_X \leq \psi(\|A^\beta u\|_X + \|A^\beta v\|_X)[\|A^\eta(u-v)\|_X$$
$$+ (\|A^\eta u\|_X + \|A^\eta v\|_X)\|A^\beta(u-v)\|_X], \qquad u, v \in \mathscr{D}(A^\eta), \tag{1.18}$$

with respect to the fractional powers A^η and A^β, where β is a second exponent such that $0 \leq \beta \leq \eta < 1$ and $\psi(\cdot) > 0$ denotes some increasing function.

Then, we have the existence and uniqueness for the strict solution to (1.17).

Theorem 1.4 *Under* (1.18), *for any* $u_0 \in \mathscr{D}(A^\beta)$, *there exists* $T_{u_0} > 0$ *such that* (1.17) *possesses a unique local solution on* $[0, T_{u_0}]$ *in the function space*

$$u \in \mathscr{C}((0, T_{u_0}]; \mathscr{D}(A)) \cap \mathscr{C}([0, T_{u_0}]; \mathscr{D}(A^\beta)) \cap \mathscr{C}^1((0, T_{u_0}]; X_{\mathbb{C}}) \qquad (1.19)$$

and that the local solution satisfies the estimate

$$t^{1-\beta} \|Au(t)\|_X + \|A^\beta u(t)\|_X \leq C_{u_0}, \qquad 0 < t \leq T_{u_0}. \qquad (1.20)$$

Here, the constant C_{u_0} *and the time* T_{u_0} *are determined depending only on the magnitude of* $\|A^\beta u_0\|_X$.

Here and in what follows, we denote by $\mathscr{C}(I; X_{\mathbb{C}})$ (resp. $\mathscr{C}^m(I; X_{\mathbb{C}})$, where $m = 1, 2, 3, \ldots$) the space of all continuous (resp. m-times continuously differentiable) functions on an interval I with values in a Banach space $X_{\mathbb{C}}$. Meanwhile, we denote by $\mathscr{B}(I; X_{\mathbb{C}})$ the space of all uniformly bounded functions on I.

1.2 Real Banach Spaces and Hilbert Spaces

We will introduce a notion of conjugation on complex Banach spaces. Owing to the conjugation, real subspaces of complex Banach spaces can be defined.

1.2.1 Conjugated Spaces

Let $X_{\mathbb{C}}$ be a complex Banach space with norm $\| \cdot \|_X$. Assume that $X_{\mathbb{C}}$ is equipped with a correspondence $f \mapsto \overline{f}$ from $X_{\mathbb{C}}$ into itself having the following properties:

$$\overline{f + g} = \overline{f} + \overline{g}, \qquad f, g \in X_{\mathbb{C}}, \qquad (1.21)$$

$$\overline{\alpha f} = \overline{\alpha}\, \overline{f}, \qquad \alpha \in \mathbb{C},\ f \in X_{\mathbb{C}}, \qquad (1.22)$$

$$\overline{\overline{f}} = f, \qquad f \in X_{\mathbb{C}}, \qquad (1.23)$$

$$\|f\|_X = \|\overline{f}\|_X, \qquad f \in X_{\mathbb{C}}. \qquad (1.24)$$

It immediately follows from these properties that this correspondence is continuous on $X_{\mathbb{C}}$ and one-to-one and onto. In particular, $\overline{0} = 0$. The vector $\overline{f}$ is called the conjugate vector of f and such a correspondence is called *a conjugation on* $X_{\mathbb{C}}$ and $X_{\mathbb{C}}$ is said to be *a conjugated space*.

For each $f \in X_{\mathbb{C}}$, we put $\operatorname{Re} f = \frac{f + \overline{f}}{2}$ and $\operatorname{Im} f = \frac{f - \overline{f}}{2i}$. Noting that $f = \operatorname{Re} f + i\operatorname{Im} f$, we call $\operatorname{Re} f$ (resp. $\operatorname{Im} f$) the real part (resp. imaginary part) of f with respect to the conjugation $f \mapsto \overline{f}$. Then, the vectors satisfying $\operatorname{Im} f = 0$ or equivalently $\overline{f} = f$ are called real vectors. By (1.21)–(1.23), both $\mathrm{Re}\,f$ and $\mathrm{Im}\,f$ are real vectors. As noticed, 0 is also a real vector. On the other hand, the vectors satisfying $\mathrm{Re}\,f = 0$ or equivalently $\overline{f} = -f$ are called purely imaginary vectors. Obviously, $i\mathrm{Im}\,f$ is a purely imaginary vector. Using $\mathrm{Re}\,f$ and $\mathrm{Im}\,f$, the conjugate vector $\overline{f}$ of f is written as $\overline{f} = \mathrm{Re}\,f - i\mathrm{Im}\,f$.

We want to consider the subset

$$X = \{f \in X_{\mathbb{C}};\ \mathrm{Im}\,f = 0, \text{i.e.}, \overline{f} = f\}. \tag{1.25}$$

By (1.21) and (1.22), X is a real linear space equipped with the norm $\|\cdot\|_X$. Therefore, X is a real Banach space, which is called *the real subspace of* $X_{\mathbb{C}}$.

Each vector $f \in X_{\mathbb{C}}$ is uniquely expressed of the form $f = \mathrm{Re}\,f + i\mathrm{Im}\,f$, where $\mathrm{Re}\,f$, $\mathrm{Im}\,f \in X$. In this sense, $X_{\mathbb{C}}$ is algebraically decomposed into the direct sum $X + iX$. In the meantime, by (1.24) it is easily seen that the correspondence $f \leftrightarrow (\mathrm{Re}\,f, \mathrm{Im}\,f)$ is bicontinuous from $X_{\mathbb{C}}$ onto $X \times X$, together with

$$\max\{\|\mathrm{Re}\,f\|_X, \|\mathrm{Im}\,f\|_X\} \le \|f\|_X \le \|\mathrm{Re}\,f\|_X + \|\mathrm{Im}\,f\|_X, \qquad f \in X_{\mathbb{C}}.$$

In this sense, we can say that $X_{\mathbb{C}}$ is also topologically decomposed into $X + iX$.

Theorem 1.5 *If $X_{\mathbb{C}}$ is a Banach space with conjugation, then the subset X given by (1.25) is a real Banach space with the norm $\|\cdot\|_X$ and $X_{\mathbb{C}}$ is algebraically and topologically decomposed into the form*

$$X_{\mathbb{C}} = X + iX.$$

Any conjugation of $X_{\mathbb{C}}$ naturally induces a conjugation on its dual space $X'_{\mathbb{C}}$.

Theorem 1.6 *If a Banach space $X_{\mathbb{C}}$ is equipped with a conjugation, then so is its dual space $X'_{\mathbb{C}}$.*

Proof Indeed, consider the correspondence $\varphi \mapsto \overline{\varphi}$ in $X'_{\mathbb{C}}$ such that $\overline{\varphi}(f) = \overline{\varphi(\overline{f})}$ or $\overline{\varphi}(\overline{f}) = \overline{\varphi(f)}$ for $f \in X_{\mathbb{C}}$. Then, it is easy to see that $\overline{\varphi} \in X'_{\mathbb{C}}$ and this correspondence fulfils all (1.21)–(1.24). $\square$

The real subspace of $X'_{\mathbb{C}}$ is denoted by X'. Therefore, we have $X'_{\mathbb{C}} = X' + iX'$.

Remark 1.2 If a Banach space $X_{\mathbb{C}}$ possesses a conjugation $f \mapsto \overline{f}$, then $\widetilde{f} = e^{i\theta}\overline{f}$, where $0 < \theta < 2\pi$, defines another conjugation on $X_{\mathbb{C}}$. So, the conjugation on $X_{\mathbb{C}}$ is not uniquely defined.

Let us now consider the case where $X_{\mathbb{C}}$ is a complex Hilbert space with inner product $(\cdot, \cdot)_X$ and where $X_{\mathbb{C}}$ is equipped with a conjugation $f \mapsto \overline{f}$. As well known, the inner product can be represented by a sum of norms as

$$4(f, g)_X = \|f + g\|_X^2 - \|f - g\|_X^2 + i\|f + ig\|_X^2 - i\|f - ig\|_X^2, \qquad f, g \in X_{\mathbb{C}}.$$

Thereby, $4\overline{(f, g)}_X = \|f+g\|_X^2 - \|f-g\|_X^2 - i\|f+ig\|_X^2 + i\|f-ig\|_X^2$. Meanwhile, we observe by (1.21) and (1.22) that

$$4(\overline{f}, \overline{g})_X = \|\overline{f} + \overline{g}\|_X^2 - \|\overline{f} - \overline{g}\|_X^2 + i\|\overline{f} + i\overline{g}\|_X^2 - i\|\overline{f} - i\overline{g}\|_X^2$$
$$= \|\overline{f+g}\|_X^2 - \|\overline{f-g}\|_X^2 + i\|\overline{f-ig}\|_X^2 - i\|\overline{f+ig}\|_X^2.$$

Then, (1.24) yields the equality

$$\overline{(f, g)}_X = (\overline{f}, \overline{g})_X, \qquad f, g \in X_{\mathbb{C}}, \tag{1.26}$$

which means that the inner product $(f, g)_X$ is real for $f, g \in X$, where X is the real subspace defined by (1.25). Thus, $(\cdot, \cdot)_X$ provides a real inner product of X.

Theorem 1.7 *If $X_{\mathbb{C}}$ is a Hilbert space with conjugation, then its real subspace X given by (1.25) possesses a structure of the real Hilbert space.*

Conversely, let $X_{\mathbb{R}}$ be a real Hilbert space with inner product $(\cdot, \cdot)_X$. We can then set a complex linear space $X_{\mathbb{C}} = X_{\mathbb{R}} + iX_{\mathbb{R}}$ in the usual manner. Moreover, for the pair of vectors of $X_{\mathbb{C}}$, define their inner product by means of

$$(f + ig, f' + ig')_X = (f, f')_X + i(g, f')_X - i(f, g')_X + (g, g')_X. \tag{1.27}$$

Then, it is not difficult to verify that this manner defines a complex inner product on $X_{\mathbb{C}}$. In addition, it holds for the norm of $X_{\mathbb{C}}$ that

$$\|f + ig\|_X^2 = (f + ig, f + ig)_X = \|f\|_X^2 + \|g\|_X^2, \qquad f + ig \in X_{\mathbb{C}}. \tag{1.28}$$

We can therefore state the converse assertion of Theorem 1.7.

Theorem 1.8 *Let a real Hilbert space $X_{\mathbb{R}}$ be given. Then, the linear space $X_{\mathbb{C}} = X_{\mathbb{R}} + iX_{\mathbb{R}}$ equipped with the inner product (1.27) is a complex Hilbert space whose real subspace is just $X_{\mathbb{R}}$.*

Proof It remains to prove the last assertion of theorem. But it is obvious if we equip a conjugation on $X_{\mathbb{C}}$ as $f + ig \mapsto f - ig$ for $f + ig \in X_{\mathbb{C}}$. It is immediate to verify that this correspondence satisfies all (1.21)–(1.24). Of course, the real subspace X of $X_{\mathbb{C}}$ given by (1.25) is nothing more than the original space $X_{\mathbb{R}}$. □

Remark 1.3 However, the situation is quite different if the given space $X_{\mathbb{R}}$ is a real Banach space. If we set the quantity $\|f\|_X$ for $f = f + ig \in X_{\mathbb{C}}$ by the same formula as (1.28), i.e., $\|f\|_X = \sqrt{\|f\|_X^2 + \|g\|_X^2}$, then $\|\cdot\|_X$ satisfies only

$$\begin{cases} \|af\|_X = |a|\|f\|_X & \text{for } a \in \mathbb{R} \text{ and } f \in X_{\mathbb{C}}, \\ \|\alpha f\|_X \le \sqrt{2}|\alpha|\|f\|_X & \text{for } \alpha \in \mathbb{C} \text{ and } f \in X_{\mathbb{C}}. \end{cases}$$

That is, $\| \cdot \|_X$ does not define a norm on $X_{\mathbb{C}}$ but only a quasi-norm (i.e., $\alpha_n \to \alpha$ in $\mathbb{C}$ and $\mathfrak{f}_n \to \mathfrak{f}$ in X imply $\alpha_n \mathfrak{f}_n \to \alpha\mathfrak{f}$ in X). A linear space equipped with a quasi-norm is called a quasi-normed space, see [Yos80, Definition 2, P. 31]. By (1.28), the converse of Theorem 1.5 cannot be proved.

1.2.2 Interpolation in Conjugated Spaces

Let $X_{\mathbb{C}}$ and $Z_{\mathbb{C}}$ be two complex Banach spaces with norm $\| \cdot \|_X$ and $\| \cdot \|_Z$ respectively such that $Z_{\mathbb{C}} \subset X_{\mathbb{C}}$ densely and continuously. Assume that both $X_{\mathbb{C}}$ and $Z_{\mathbb{C}}$ are equipped with conjugation. In addition, we assume that these conjugations are consistent in the sense that the conjugation $f \mapsto \overline{f}$ on $X_{\mathbb{C}}$ enjoys the properties:

$$u \in Z_{\mathbb{C}} \quad \text{if and only if} \quad \overline{u} \in Z_{\mathbb{C}}, \tag{1.29}$$

$$\|u\|_Z = \|\overline{u}\|_Z, \qquad u \in Z_{\mathbb{C}}. \tag{1.30}$$

For $0 \le \theta \le 1$, let $[X_{\mathbb{C}}, Z_{\mathbb{C}}]_\theta$ be the interpolation spaces reviewed in Sect. 1.3. Then, the consistent conjugation on $X_{\mathbb{C}}$ and on $Z_{\mathbb{C}}$ runs to those spaces.

Proposition 1.1 *For any $0 \le \theta \le 1$, it holds true that*

$$u \in [X_{\mathbb{C}}, Z_{\mathbb{C}}]_\theta \quad \text{if and only if} \quad \overline{u} \in [X_{\mathbb{C}}, Z_{\mathbb{C}}]_\theta, \tag{1.31}$$

$$\|u\|_\theta = \|\overline{u}\|_\theta, \qquad u \in [X_{\mathbb{C}}, Z_{\mathbb{C}}]_\theta. \tag{1.32}$$

Proof Let $u \in [X_{\mathbb{C}}, Z_{\mathbb{C}}]_\theta$. By definition, there exists an analytic function $F(z)$ in the function space $\mathscr{H}(X_{\mathbb{C}}, Z_{\mathbb{C}})$ introduced above for which it holds that $F(\theta) = u$. Then, due to (1.22), the function $G(z) = \overline{F(\overline{z})}$ also belongs to $\mathscr{H}(X_{\mathbb{C}}, Z_{\mathbb{C}})$ and takes a value $G(\theta) = \overline{u}$ at $z = \theta$. This then means that $\overline{u}$ is in $[X_{\mathbb{C}}, Z_{\mathbb{C}}]_\theta$. Conversely, if $\overline{u} \in [X_{\mathbb{C}}, Z_{\mathbb{C}}]_\theta$, then $u = \overline{\overline{u}} \in [X_{\mathbb{C}}, Z_{\mathbb{C}}]_\theta$. Hence, (1.31) is verified. The equality (1.32) is also easily verified from the definition (1.5) of a norm. □

This proposition shows that the conjugation on $X_{\mathbb{C}}$ satisfying (1.29)–(1.30) always induces a conjugation on the interpolation space $[X_{\mathbb{C}}, Z_{\mathbb{C}}]_\theta$, too, which is consistent with the original one. Then, consider the real subspace $([X_{\mathbb{C}}, Z_{\mathbb{C}}]_\theta)_{\mathbb{R}}$ of $[X_{\mathbb{C}}, Z_{\mathbb{C}}]_\theta$ defined by (1.25). Theorem 1.5 yields the following theorem.

Theorem 1.9 *The subspace $([X_{\mathbb{C}}, Z_{\mathbb{C}}]_\theta)_{\mathbb{R}}$ is a real Banach space with the norm $\| \cdot \|_\theta$ and $[X_{\mathbb{C}}, Z_{\mathbb{C}}]_\theta$ is algebraically and topologically decomposed into*

$$[X_{\mathbb{C}}, Z_{\mathbb{C}}]_\theta = ([X_{\mathbb{C}}, Z_{\mathbb{C}}]_\theta)_{\mathbb{R}} + i\,([X_{\mathbb{C}}, Z_{\mathbb{C}}]_\theta)_{\mathbb{R}}, \qquad 0 < \theta < 1. \tag{1.33}$$

In this way, we are led to introduce a definition of interpolation for real spaces.

Definition 1.1 Let two complex Banach spaces $Z_\mathbb{C} \subset X_\mathbb{C}$ densely and continuously be equipped with a consistent conjugation. Let X and Z be the real subspaces of $X_\mathbb{C}$ and $Z_\mathbb{C}$ respectively. Then, their interpolation space $[X, Z]_\theta$ is defined by

$$[X, Z]_\theta = ([X_\mathbb{C}, Z_\mathbb{C}]_\theta)_\mathbb{R}, \qquad 0 \leq \theta \leq 1. \qquad (1.34)$$

1.2.3 Triplets of Conjugated Spaces

Let $X_\mathbb{C}$ be a complex Hilbert space with inner product $(\cdot, \cdot)_X$ and norm $\| \cdot \|_X$ and let $Z_\mathbb{C}$ be a second complex Hilbert space with norm $\| \cdot \|_Z$. Assume that $Z_\mathbb{C} \subset X_\mathbb{C}$ densely and continuously. Then, as reviewed in Subsection 1.4, we have a triplet $Z_\mathbb{C} \subset X_\mathbb{C} \subset Z_\mathbb{C}^*$ with the properties (1)–(3). In addition, assume that $X_\mathbb{C}$ and $Z_\mathbb{C}$ are equipped with a consistent conjugation in the sense of (1.29)–(1.30). Then, we obtain the following results.

First, we see that the conjugation runs to the third space $Z_\mathbb{C}^*$. Indeed, on account of (1.8), (1.9), (1.26) and (1.30), it holds for $f \in X_\mathbb{C}$ that

$$\|f\|_{Z^*} = \sup_{\|u\|_Z \leq 1} |(u, f)_X| = \sup_{\|u\|_Z \leq 1} |\overline{(u, f)_X}|$$

$$= \sup_{\|u\|_Z \leq 1} |(\overline{u}, \overline{f})_X| = \sup_{\|\overline{u}\|_Z \leq 1} |(\overline{u}, \overline{f})_X| = \|\overline{f}\|_{Z^*}.$$

This means that $f \mapsto \overline{f}$ is continuous also with respect to the norm $\| \cdot \|_{Z^*}$, on $X_\mathbb{C}$. Since $X_\mathbb{C}$ is dense in $Z_\mathbb{C}^*$, it is possible to extend the conjugation $f \mapsto \overline{f}$ continuously over $Z_\mathbb{C}^*$ with the equality $\|\varphi\|_{Z^*} = \|\overline{\varphi}\|_{Z^*}$ for all $\varphi \in Z_\mathbb{C}^*$. Similarly, it follows from (1.9) and (1.26) that

$$\overline{\langle u, \varphi \rangle}_{Z \times Z^*} = \langle \overline{u}, \overline{\varphi} \rangle_{Z \times Z^*}, \qquad u \in Z_\mathbb{C}, \ \varphi \in Z_\mathbb{C}^*. \qquad (1.35)$$

Let us denote the real subspaces of $Z_\mathbb{C}$, $X_\mathbb{C}$ and $Z_\mathbb{C}^*$ by Z, X and Z^*, respectively. It is then clear that $Z \subset X \subset Z^*$ densely and continuously. The property (1.35) means that the duality product $\langle \cdot, \cdot \rangle_{Z \times Z^*}$ is real on $Z \times Z^*$.

Second, we see that

$$\|u\|_Z = \sup_{\|\varphi\|_{Z^*} \leq 1, \, \varphi \in Z_\mathbb{R}^*} |\langle u, \varphi \rangle_{Z \times Z^*}| \qquad \text{for } u \in Z, \qquad (1.36)$$

$$\|\varphi\|_{Z^*} = \sup_{\|u\| \leq 1, \, u \in Z_\mathbb{R}} |\langle u, \varphi \rangle_{Z \times Z^*}| \qquad \text{for } \varphi \in Z^*. \qquad (1.37)$$

Indeed, by Theorem 1.2, there exists for $u \in Z$ an element $\varphi \in Z_\mathbb{C}^*$ such that $\|\varphi\|_{Z^*} = 1$ and $\|u\|_Z = \langle u, \varphi \rangle_{Z \times Z^*}$. On the other hand, by (1.35) it follows that $\|u\|_Z = \overline{\langle u, \varphi \rangle}_{Z \times Z^*} = \langle u, \overline{\varphi} \rangle_{Z \times Z^*}$. Therefore, we obtain that $\|u\|_Z = \langle u, \text{Re } \varphi \rangle_{Z \times Z^*}$ together with $\|\text{Re}\varphi\|_{Z^*} \le \|\varphi\|_{Z^*} \le 1$. Hence, (1.36) is proved.

For $\varphi \in Z^*$, there exists a sequence $u_n \in Z_{\mathbb{C}}$ such that $\|u_n\|_Z \le 1$ and $\langle u_n, \varphi \rangle_{Z \times Z^*} \to \|\varphi\|_{Z^*}$. By (1.35), it is the same for the sequence $\overline{u}_n$. Then, $\langle \text{Re } u_n, \varphi \rangle_{Z \times Z^*} \to \|\varphi\|_{Z^*}$ together with $\|\text{Re } u_n\|_Z \le \|u_n\|_Z \le 1$. Hence, (1.37) is also proved.

Third, let Z', X' and $(Z^*)'$ be the real subspaces of the dual spaces $Z'_{\mathbb{C}}$, $X'_{\mathbb{C}}$ and $(Z^*_{\mathbb{C}})'$, respectively, obtained by Theorem 1.6. Then, we see that

$$\begin{cases} \text{the isometric isomorphism } J : Z^*_{\mathbb{C}} \to Z'_{\mathbb{C}} \text{ maps } Z^* \text{ onto } Z', \\ \text{the isometric isomorphism } J : X_{\mathbb{C}} \to X'_{\mathbb{C}} \text{ maps } X \text{ onto } X', \\ \text{the isometric isomorphism } J : Z_{\mathbb{C}} \to (Z^*_{\mathbb{C}})' \text{ maps } Z \text{ onto } (Z^*)'. \end{cases} \qquad (1.38)$$

In fact, let $\varphi \in Z^*$, i.e., $\varphi = \overline{\varphi}$,; then, by Theorem 1.6, $[\overline{J^*\varphi}](u) = \overline{[J^*\varphi](\overline{u})} = \overline{\langle \overline{u}, \varphi \rangle}_{Z \times Z^*}$ for $u \in Z_{\mathbb{C}}$; therefore, by (1.35), $[\overline{J^*\varphi}](u) = \langle u, \varphi \rangle_{Z \times Z^*} = [J^*\varphi](u)$; hence, $\overline{J^*\varphi} = J\varphi$, i.e., $J^*\varphi \in Z'$. Conversely, let $J^*\varphi \in Z'$; on one hand, $[J^*\varphi](u) = \langle u, \varphi \rangle_{Z \times Z^*}$ for $u \in Z_{\mathbb{C}}$; on the other hand, as before, $[\overline{J^*\varphi}](u) = \overline{\langle \overline{u}, \varphi \rangle}_{Z \times Z^*} = \langle u, \overline{\varphi} \rangle_{Z \times Z^*}$ for $u \in Z_{\mathbb{C}}$; hence, $\varphi = \overline{\varphi}$, i.e., $\varphi \in Z^*$. Thus, the first assertion of (1.38) has been observed.

Analogously, the second and third assertions of (1.38) can be observed by using Theorems 1.1 and 1.2, (1.26) and (1.35).

Finally, by the definition (1.34), the fact (1.10) yields the interpolation property $[Z^*, Z]_{\frac{1}{2}} = X$.

We have thus arrived at the following results.

Theorem 1.10 *Let $Z_{\mathbb{C}} \subset X_{\mathbb{C}} \subset Z^*_{\mathbb{C}}$ be a triplet of complex spaces. Assume that $X_{\mathbb{C}}$ and $Z_{\mathbb{C}}$ are equipped with a consistent conjugation. Then, the conjugation naturally runs to $Z^*_{\mathbb{C}}$, which is still consistent with the original one. Furthermore, the real subspaces Z, X and Z^* of $Z_{\mathbb{C}}$, $X_{\mathbb{C}}$ and $Z^*_{\mathbb{C}}$, respectively, enjoy the following properties:*

1. *$Z \subset X \subset Z^*$ with dense and continuous embeddings.*
2. *$\{Z, Z^*\}$ forms an adjoint pair with a real duality product $\langle \cdot, \cdot \rangle_{Z \times Z^*}$ satisfying (1.36) and (1.37) instead of (1.7) and (1.8) respectively.*
3. *The isometric isomorphism J has the properties in (1.38).*
4. *The property $[Z^*, Z]_{\frac{1}{2}} = X$ holds true.*

Definition 1.2 Let $Z_{\mathbb{C}} \subset X_{\mathbb{C}} \subset Z^*_{\mathbb{C}}$ be a triplet of complex spaces with consistent conjugation as above. Let Z, X and Z^* be real subspaces of $Z_{\mathbb{C}}$, $X_{\mathbb{C}}$ and $Z^*_{\mathbb{C}}$, respectively. Then, we will call $Z \subset X \subset Z^*$ having the properties announced in Theorem 1.10 *a triplet of real spaces.*

Remark 1.4 Now, let a real Hilbert space $X_{\mathbb{R}}$ be given with inner product $(\cdot, \cdot)_X$ and norm $\| \cdot \|_X$ and let a second real Hilbert space $Z_{\mathbb{R}}$ be given with norm $\| \cdot \|_Z$. Assume that $Z_{\mathbb{R}} \subset X_{\mathbb{R}}$ densely and continuously. As shown by Theorem 1.8, there exists their complexification to complex Hilbert spaces $X_{\mathbb{C}} = X_{\mathbb{R}} + iX_{\mathbb{R}}$ and

$Z_{\mathbb{C}} = Z_{\mathbb{R}} + iZ_{\mathbb{R}}$ respectively. It is clear that $Z_{\mathbb{C}} \subset X_{\mathbb{C}}$ densely and continuously; consequently, we have a triplet $Z_{\mathbb{C}} \subset X_{\mathbb{C}} \subset Z_{\mathbb{C}}^*$. Furthermore, it is also clear that $f + ig \mapsto f - ig$ for $f + ig \in X_{\mathbb{C}}$ defines a consistent conjugation on $X_{\mathbb{C}}$ and on $Z_{\mathbb{C}}$. Therefore, all the results obtained by Theorem 1.10 are available. In other words, if two real Hilbert spaces $X_{\mathbb{R}}$ and $Z_{\mathbb{R}}$ are given as above, then we can always construct a triplet $Z \subset X \subset Z^*$ of real spaces, where $X = X_{\mathbb{R}}$ and $Z = Z_{\mathbb{R}}$.

In fact, the triplet $Z \subset X \subset Z^*$ will give us a general framework of our study. A function $u(t)$ will be given in the function space $u \in \mathscr{C}([0, \infty); Z) \cap \mathscr{C}^1([0, \infty); Z^*)$ and there will exist a suitable function $\Phi : X \to \mathbb{R}$ for which the value $\Phi(u(t))$ is monotonously decreasing for $0 \le t < \infty$. Under these situations, we are going to discuss convergence of $u(t)$ as $t \to \infty$.

1.2.4 Real Sobolev–Lebesgue Spaces

Let Ω be an n-dimensional bounded domain with Lipschitz boundary. For $1 \le p \le \infty$, let $L_p(\Omega; \mathbb{C})$ denote the Banach space consisting of all complex L_p-functions in Ω equipped with the usual L_p-norm. Clearly, the correspondence $f(x) \mapsto \overline{f}(x) \equiv \overline{f(x)}$ gives a conjugation on $L_p(\Omega; \mathbb{C})$ (thanks to the conjugation of $\mathbb{C}$ in spite of Remark 1.3). Then, we see that

$$L_p(\Omega; \mathbb{C}) = L_p(\Omega; \mathbb{R}) + iL_p(\Omega; \mathbb{R})$$

in the sense of Theorem 1.5, where $L_p(\Omega; \mathbb{R})$ is the space of real L_p-functions in Ω.

For $1 < p < \infty$ and integral $m = 0, 1, 2, \ldots$, the complex Sobolev space $H_p^m(\Omega; C)$ of order m is characterized by

$$H_p^m(\Omega; \mathbb{C}) = \{u \in L_p(\Omega; \mathbb{C}); \ D^\alpha u \in L_p(\Omega; \mathbb{C}) \text{ for all } \alpha \text{ up to } |\alpha| \le m\},$$

here $\alpha = (\alpha_1, \alpha_2, \ldots, \alpha_n)$ is a multi-index, $|\alpha|$ stands for $\sum_{i=1}^n \alpha_i$, and $D^\alpha u = D_{x_1}^{\alpha_1} D_{x_2}^{\alpha_2} \cdots D_{x_n}^{\alpha_n} u$ is the derivative of u in the distribution sense. Its norm is given by $\|u\|_{H_p^m} = \sum_{0 \le |\alpha| \le m} \|D^\alpha u\|_{L_p}$. Clearly, the conjugation on $L_p(\Omega; \mathbb{C})$ is consistent with that on $H_p^m(\Omega; \mathbb{C})$. As for $L_p(\Omega; \mathbb{C})$, we have

$$H_p^m(\Omega; \mathbb{C}) = H_p^m(\Omega; \mathbb{R}) + iH_p^m(\Omega; \mathbb{R})$$

in the sense of Theorem 1.5, where $H_p^m(\Omega; \mathbb{R})$ is the real Sobolev space of order m.

For $1 < p < \infty$ and fractional s such that $m < s < m+1$, the complex Lebesgue space $H_p^s(\Omega; \mathbb{C})$ of order s is known to be characterized as

$$H_p^s(\Omega; \mathbb{C}) = [H_p^m(\Omega; \mathbb{C}), H_p^{m+1}(\Omega; \mathbb{C})]_\theta, \qquad \theta = s - m.$$

Then, Theorem 1.9 is available.

It follows by (1.33) that $H_p^s(\Omega; \mathbb{C}) = H_p^s(\Omega; \mathbb{R}) + i H_p^s(\Omega; \mathbb{R})$ algebraically and topologically, where $H_p^s(\Omega; \mathbb{R})$ is the real Lebesgue space of order s. In particular, we know that $u \in H_p^s(\Omega; \mathbb{C})$ if and only if $\mathrm{Re}\,u$, $\mathrm{Im}\,u \in H_p^s(\Omega; \mathbb{R})$ and it holds that

$$\max\{\|\mathrm{Re}\,u\|_{H_p^s}, \|\mathrm{Im}\,u\|_{H_p^s}\} \le \|u\|_{H_p^s} \le \|\mathrm{Re}\,u\|_{H_p^s} + \|\mathrm{Im}\,u\|_{H_p^s}, \quad u \in H_p^s(\Omega; \mathbb{C}).$$

Furthermore, according to Definition 1.1, we have

$$H_p^s(\Omega; \mathbb{R}) = [H_p^m(\Omega; \mathbb{R}), H_p^{m+1}(\Omega; \mathbb{R})]_\theta, \qquad \theta = s - m.$$

1.3 Real Sectorial Operators

In conjugated complex Banach spaces, we introduce a notion of real linear operators acting in the real subspaces, furthermore, that of real sectorial operators.

1.3.1 Real Operators in Conjugated Spaces

Let $X_\mathbb{C}$ be a complex Banach space with norm $\|\cdot\|_X$ and with conjugation $f \mapsto \overline{f}$, X being its real subspace. Let A denote a complex linear operator from $\mathscr{D}(A) \subset X_\mathbb{C}$ into $X_\mathbb{C}$.

Assume that an operator A satisfies the following two conditions:

$$u \in \mathscr{D}(A) \quad \text{if and only if} \quad \overline{u} \in \mathscr{D}(A); \tag{1.39}$$

$$\overline{Au} = A\overline{u} \qquad \text{for } u \in \mathscr{D}(A). \tag{1.40}$$

Following (1.25), let us define

$$[\mathscr{D}(A)]_\mathbb{R} = \{u \in \mathscr{D}(A); \ u = \overline{u}, \text{i.e., } \mathrm{Im}\,u = 0\} \subset X.$$

Then, (1.39) implies that $[\mathscr{D}(A)]_\mathbb{R}$ is a real linear subspace of X and that $\mathscr{D}(A)$ is decomposed algebraically into

$$\mathscr{D}(A) = [\mathscr{D}(A)]_\mathbb{R} + i[\mathscr{D}(A)]_\mathbb{R}.$$

In addition, (1.40) implies that, if $u \in [\mathscr{D}(A)]_\mathbb{R}$, then $\overline{Au} = A\overline{u} = Au$ i.e., $Au \in X$, which shows that A maps $[\mathscr{D}(A)]_\mathbb{R}$ into X.

Therefore, we notice the following fact.

Theorem 1.11 *Let a complex linear operator A satisfy* (1.39)–(1.40). *Then, $[\mathscr{D}(A)]_\mathbb{R}$ is a real linear subspace of X such that $\mathscr{D}(A) = [\mathscr{D}(A)]_\mathbb{R} + i[\mathscr{D}(A)]_\mathbb{R}$ algebraically, and A is a real linear operator from $[\mathscr{D}(A)]_\mathbb{R}$ into X.*

So, when a complex linear operator A enjoys the properties (1.39)–(1.40), A is called *a real operator of $X_\mathbb{C}$*. When we consider A as a real linear operator acting in X, we will denote it by the same notation A if there is no fear of confusion.

If A is a densely defined, closed linear operator of $X_\mathbb{C}$ which is real, then A is, at the same time, a densely defined, closed linear operator of X with the domain $\mathscr{D}(A) \cap X = [\mathscr{D}(A)]_\mathbb{R}$.

1.3.2 Sectorial Operators in Conjugated Spaces

Now, consider a sectorial operator A of $X_\mathbb{C}$, namely, let A be a densely defined, closed linear operator of $X_\mathbb{C}$ satisfying (1.11)–(1.12). When A is, at the same time, a real operator of $X_\mathbb{C}$, we say that A is *a real sectorial operator of $X_\mathbb{C}$*.

The resolvent of a real sectorial operator has the following properties.

Proposition 1.2 *Let A be a real sectorial operator of $X_\mathbb{C}$. Then,*

$$\lambda \in \rho(A) \quad \text{if and only if} \quad \overline{\lambda} \in \rho(A), \tag{1.41}$$

$$\overline{(\lambda - A)^{-1} f} = (\overline{\lambda} - A)^{-1} \overline{f} \quad \text{for } \lambda \in \rho(A), \ f \in X_\mathbb{C}. \tag{1.42}$$

Proof It follows from (1.39) and (1.40) that, for $u \in \mathscr{D}(A)$ and $f \in X_\mathbb{C}$, the relation $(\lambda - A)u = f$ is equivalent to $(\overline{\lambda} - A)\overline{u} = \overline{f}$. Then, $\lambda \in \rho(A)$ implies that $\overline{\lambda} - A$ is a bijection from $\mathscr{D}(A)$ onto $X_\mathbb{C}$; hence, $\overline{\lambda}$ belongs to $\rho(A)$. Similarly, $\overline{\lambda} \in \rho(A)$ implies that $\lambda \in \rho(A)$.

As seen, we have $(\lambda - A)u = f$ and $(\overline{\lambda} - A)\overline{u} = \overline{f}$ for $u \in \mathscr{D}(A)$. Therefore, when $\lambda, \overline{\lambda} \in \rho(A)$, we obtain that $u = (\lambda - A)^{-1} f$ and $\overline{u} = (\overline{\lambda} - A)^{-1} \overline{f}$. □

By properties (1.41)–(1.42) observed above for a real sectorial operator A, we can verify that its functional calculus also enjoys the properties (1.39)–(1.40).

(I) Fractional Powers Let A be a real sectorial operator of $X_\mathbb{C}$. According to (1.13), for $0 < x < \infty$, $A^{-x} f$ is given by the integral

$$A^{-x} f = \frac{1}{2\pi i} \int_\Gamma \lambda^{-x} (\lambda - A)^{-1} f d\lambda, \qquad f \in X_\mathbb{C},$$

in the space $X_\mathbb{C}$. Taking a conjugation of this equality, we obtain by (1.42) that

$$\overline{A^{-x} f} = -\frac{1}{2\pi i} \int_\Gamma (\overline{\lambda})^{-x} (\overline{\lambda} - A)^{-1} \overline{f} d\overline{\lambda}.$$

Here, we notice the fact that, when the integral contour Γ is symmetric with respect to the real axis as in (1.13), the transform $\lambda \in \Gamma \mapsto \overline{\lambda} \in \Gamma$ yields an integration along the same contour but in the opposite orientation, i.e., along $-\Gamma$. Then,

$$\overline{A^{-x}f} = -\frac{1}{2\pi i} \int_{-\Gamma} \lambda^{-x}(\lambda - A)^{-1}\overline{f}d\lambda = A^{-x}\overline{f}.$$

Hence, A^{-x} satisfies (1.40) and A^{-x} is a real bounded operator of $X_{\mathbb{C}}$.

Since $u \in \mathscr{R}(A^{-x})$ if and only if $\overline{u} \in \mathscr{R}(A^{-x})$, we see that $u \in \mathscr{D}(A^x)$ if and only if $\overline{u} \in \mathscr{D}(A^x)$. In addition, it holds that $\overline{A^x u} = A^x \overline{u}$ for $u \in \mathscr{D}(A^x)$. By the definition (1.39)–(1.40), A^x is deduced to be a real operator of $X_{\mathbb{C}}$. Consequently,

A^x is a densely defined, closed linear operator of X

with the domain $\mathscr{D}(A^x) \cap X = [\mathscr{D}(A^x)]_{\mathbb{R}}$.   (1.43)

We shall see that the fractional powers of real sectorial operators are very effective in setting up the triplets of real spaces in applications.

In view of (1.33) and (1.43), we have a real version of (1.15).

Theorem 1.12 *Let $X_{\mathbb{C}}$ be a conjugated Hilbert space and let A be a positive definite self-adjoint operator of $X_{\mathbb{C}}$ which is real in $X_{\mathbb{C}}$. Let $0 \leq \theta_0 < \theta_1 \leq 1$. Then, for $\theta_0 < \theta < \theta_1$, it holds that*

$$[\mathscr{D}(A^\theta)]_{\mathbb{R}} = \left([\mathscr{D}(A^{\theta_0}), \mathscr{D}(A^{\theta_1})]_{\theta'}\right)_{\mathbb{R}}, \quad where \ (1-\theta')\theta_0 + \theta'\theta_1 = \theta. \quad (1.44)$$

(II) Analytic Semigroups Let A be a real sectorial operator of $X_{\mathbb{C}}$ whose angle is such that $\omega_A < \frac{\pi}{2}$. By the same arguments as for the fractional powers A^{-x}, the semigroup e^{-tA} which is given by the integral (1.16) is seen to satisfy the condition (1.40) and to be a real operator of $X_{\mathbb{C}}$. Consequently, e^{-tA} is a bounded linear operator acting on X and

e^{-tA} defines an analytic semigroup on X.   (1.45)

1.4 Operators Associated with Sesquilinear Forms

As is well known, making use of sesquilinear forms in the framework of triplets of spaces is the most convenient manner for determining sectorial operators (see Dautray–Lions [DL84b, Chapitre VII]). These techniques are still very convenient for determining various real sectorial operators.

1.4.1 Real Sesquilinear Forms

Let $X_{\mathbb{C}}$ be a complex Hilbert space with inner product $(\cdot, \cdot)_X$ and norm $\| \cdot \|_X$ and let $Z_{\mathbb{C}}$ be a second complex Hilbert space with norm $\| \cdot \|_Z$, $Z_{\mathbb{C}}$ being embedded in $X_{\mathbb{C}}$ densely and continuously. Let $Z_{\mathbb{C}} \subset X_{\mathbb{C}} \subset Z_{\mathbb{C}}^*$ be the triplet of complex spaces.

Consider a sesquilinear form $a(u, v)$ defined on $Z_{\mathbb{C}} \times Z_{\mathbb{C}}$ (see (1.3)). Assume that $a(u, v)$ satisfies the conditions:

$$\begin{cases} |a(u, v)| \leq M \|u\|_Z \|v\|_Z & \text{for } u, v \in Z_{\mathbb{C}}, \\ \operatorname{Re} a(u, u) \geq \delta \|u\|_Z^2 & \text{for } u \in Z_{\mathbb{C}}, \end{cases} \quad (1.46)$$

with some constants $M > 0$ and $\delta > 0$.

For each fixed $u \in Z_{\mathbb{C}}$, the correspondence $v \mapsto \overline{a(u, v)}$ is a bounded linear functional. Then, by virtue of Theorem 1.2, there exists a vector $\mathscr{A}u \in Z_{\mathbb{C}}^*$ determined from u such that $\overline{a(u, v)} = \langle v, \mathscr{A}u \rangle_{Z \times Z^*}$ for all $v \in Z_{\mathbb{C}}$, i.e., $a(u, v) = \langle \mathscr{A}u, v \rangle_{Z^* \times Z}$, $\forall v \in Z_{\mathbb{C}}$. Under (1.46), the correspondence $u \mapsto \mathscr{A}u$ is seen to be a linear isomorphism from $Z_{\mathbb{C}}$ onto $Z_{\mathbb{C}}^*$. Furthermore, when $\mathscr{A}$ is considered as a closed linear operator acting in $Z_{\mathbb{C}}^*$, $\mathscr{A}$ is shown to be a sectorial operator of angle $\omega_{\mathscr{A}} < \frac{\pi}{2}$. In addition, its part $A = \mathscr{A}_{|X_{\mathbb{C}}}$ in $X_{\mathbb{C}}$, which is defined by the domain $\mathscr{D}(A) = \{u \in Z_{\mathbb{C}}; \ \mathscr{A}u \in X_{\mathbb{C}}\}$, is also shown to be a densely defined, closed linear operator and moreover to be a sectorial operator of $X_{\mathbb{C}}$ of angle $\omega_A < \frac{\pi}{2}$.

Now, assume that a consistent conjugation $f \mapsto \overline{f}$ is equipped on $X_{\mathbb{C}}$ and on $Z_{\mathbb{C}}$ (see (1.29)–(1.30)). As noticed in Theorem 1.10, this conjugation runs to that on $Z_{\mathbb{C}}^*$ which is still consistent. Let $Z \subset X \subset Z^*$ be the triplet of real spaces (see Definition 1.2).

We want to consider a sesquilinear form $a(u, v)$ on $Z_{\mathbb{C}} \times Z_{\mathbb{C}}$ satisfying

$$\overline{a(u, v)} = a(\overline{u}, \overline{v}) \quad \text{for} \quad u, v \in Z_{\mathbb{C}}. \quad (1.47)$$

This condition obviously implies that $a(u, v)$ is real if u and v are real in $Z_{\mathbb{C}}$, i.e., $u, v \in Z$. In this sense, a sesquilinear form satisfying (1.47) is called *a real sesquilinear form*.

By the property (1.35), it follows from (1.47) that

$$\langle \mathscr{A}u, v \rangle_{Z^* \times Z} = \overline{\langle \mathscr{A}u, \overline{v} \rangle_{Z^* \times Z}} = \overline{\overline{a}(u, \overline{v})} = a(\overline{u}, v) = \langle \mathscr{A}\overline{u}, v \rangle_{Z^* \times Z},$$

which yields the property $\overline{\mathscr{A}u} = \mathscr{A}\overline{u}$ for $u \in Z_{\mathbb{C}}$. The definition (1.39)–(1.40) then shows that $\mathscr{A}$ is a real operator of $Z_{\mathbb{C}}^*$. In particular, $\mathscr{A}$ defines a real linear operator from Z onto Z^*.

Similarly, A is also verified to be a real operator of $X_{\mathbb{C}}$. Indeed, $u \in \mathscr{D}(A)$ if and only if $\mathscr{A}u \in X_{\mathbb{C}}$; then, since $\overline{\mathscr{A}u} = \mathscr{A}\overline{u}$, this is equivalent to $\mathscr{A}\overline{u} \in X_{\mathbb{C}}$; hence, $u \in \mathscr{D}(A)$ if and only if $\overline{u} \in D(A)$. Hence, A defines a real linear operator from $[\mathscr{D}(A)]_{\mathbb{R}}$ onto X.

Theorem 1.13 *Let $Z_{\mathbb{C}} \subset X_{\mathbb{C}} \subset Z_{\mathbb{C}}^*$ be a triplet equipped with consistent conjugation. Consider a real sesquilinear form $a(u, v)$ on $Z_{\mathbb{C}} \times Z_{\mathbb{C}}$. Then, the associated linear operator $\mathscr{A}$ (resp. A) to the form $a(u, v)$ is a real sectorial operator of $Z_{\mathbb{C}}$ (resp. $X_{\mathbb{C}}$) of angle $\omega_{\mathscr{A}} < \frac{\pi}{2}$ (resp. $\omega_A < \frac{\pi}{2}$).*

In addition to (1.46) and (1.47), assume that $a(u, v)$ is Hermitian, i.e.,

$$a(v, u) = \overline{a(u, v)}, \qquad u, v \in Z_{\mathbb{C}}. \tag{1.48}$$

This implies that $\langle \mathscr{A}u, v \rangle_{Z^* \times Z} = a(u, v) = \overline{a(v, u)} = \overline{\langle \mathscr{A}v, u \rangle_{Z^* \times Z}} = \langle u, \mathscr{A}v \rangle_{Z \times Z^*}$ for $u, v \in Z_{\mathbb{C}}$, that is, $\mathscr{A}$ is symmetric. Similarly, (1.48) implies that $(Au, v)_X = (u, Av)_X$ for $u, v \in \mathscr{D}(A)$, that is, A is a self-adjoint operator of $X_{\mathbb{C}}$. In this case, it is known that the domain of the square root of $\mathscr{A}$ (resp. A) is characterized as $\mathscr{D}(\mathscr{A}^{\frac{1}{2}}) = X_{\mathbb{C}}$ (resp. $\mathscr{D}(A^{\frac{1}{2}}) = Z_{\mathbb{C}}$). Then, due to (1.43), we observe that

$$\mathscr{D}(\mathscr{A}^{\frac{1}{2}}) \cap Z^* = [D(\mathscr{A}^{\frac{1}{2}})]_{\mathbb{R}} = X \quad \text{and} \quad \mathscr{D}(A^{\frac{1}{2}}) \cap X = [D(A^{\frac{1}{2}})]_{\mathbb{R}} = Z. \tag{1.49}$$

1.4.2 Realization of Elliptic Operators

It is possible to present by using Theorem 1.13 some examples of real sectorial operators acting in suitable real function spaces.

We denote by Ω a bounded domain of $\mathbb{R}^n$ with Lipschitz boundary. Let $L_2(\Omega; \mathbb{C})$ be the complex L_2-space in Ω with norm $\|\cdot\|_{L_2}$ and let $\mathring{H}^1(\Omega; \mathbb{C})$ denote the closure of the space $\mathscr{C}_0^\infty(\Omega; \mathbb{C})$ with respect to the Sobolev norm $\|\cdot\|_{H^1}$, where $\mathscr{C}_0^\infty(\Omega; \mathbb{C})$ is the space of test functions consisting of all $\mathscr{C}^\infty$-functions with compact support. Actually, $\mathring{H}^1(\Omega; \mathbb{C})$ is characterized as a closed subspace of $H^1(\Omega; \mathbb{C})$ consisting of functions satisfying the homogeneous Dirichlet conditions $u_{|\partial\Omega} = 0$ on the boundary $\partial\Omega$ of Ω.

As well known, $\mathring{H}^1(\Omega; \mathbb{C}) \subset L_2(\Omega; \mathbb{C}) \subset H^{-1}(\Omega; \mathbb{C})$ make a triplet, where $H^{-1}(\Omega; \mathbb{C})(\subset \mathscr{C}_0^\infty(\Omega; \mathbb{C})')$ is the adjoint space of $\mathring{H}^1(\Omega; \mathbb{C})$ with the usual duality product in the sense of distributions.

Let $f \mapsto \overline{f}$ be the complex conjugation on $L_2(\Omega; \mathbb{C})$ which obviously satisfies (1.21)–(1.24) and is consistent with the conjugation on $\mathring{H}^1(\Omega; \mathbb{C})$. Thereby, this also runs to $H^{-1}(\Omega; \mathbb{C})$. Consequently, the three complex spaces are decomposed into the direct sums of $\mathring{H}^1(\Omega; \mathbb{C}) = \mathring{H}^1(\Omega; \mathbb{R}) + i\mathring{H}^1(\Omega; \mathbb{R})$, $L_2(\Omega; \mathbb{C}) = L_2(\Omega; \mathbb{R}) + iL_2(\Omega; \mathbb{R})$ and $H^{-1}(\Omega; \mathbb{C}) = H^{-1}(\Omega; \mathbb{R}) + iH^{-1}(\Omega; \mathbb{R})$, respectively. As shown by Theorem 1.10, we have a triplet $\mathring{H}^1(\Omega; \mathbb{R}) \subset L_2(\Omega; \mathbb{R}) \subset H^{-1}(\Omega; \mathbb{R})$ of real spaces.

Consider a sesquilinear form

$$a(u, v) = \sum_{j,k=1}^{n} \int_{\Omega} a_{jk}(x) \frac{\partial u}{\partial x_j} \frac{\partial \overline{v}}{\partial x_k} dx, \qquad u, v \in \overset{\circ}{H}^1(\Omega; \mathbb{C}).$$

We here assume that the functions $a_{jk}(x)$ satisfy the conditions:

$$a_{jk} = a_{kj} \in L_\infty(\Omega; \mathbb{R}) \qquad \text{for } 1 \le j, k \le n, \tag{1.50}$$

$$\sum_{j,k=1}^{n} a_{jk}(x)\xi_j\xi_k \ge \delta|\xi|^2 \quad \text{for a.e. } x \in \Omega \text{ and } \forall \xi = (\xi_1, \ldots, \xi_n) \in \mathbb{R}^n, \tag{1.51}$$

with some constant $\delta > 0$.

By (1.50)–(1.51), the form $a(u, v)$ is verified to satisfy (1.46). The assumption (1.50) yields also the conditions (1.47) and (1.48) for $a(u, v)$. Therefore, all the results stated above are available to this real sesquilinear form $a(u, v)$.

Indeed, let $\mathscr{A}$ be the operator associated with $a(u, v)$. Then, $\mathscr{A}$ is a real sectorial operator of $H^{-1}(\Omega; \mathbb{C})$ with domain $\mathscr{D}(\mathscr{A}) = \overset{\circ}{H}^1(\Omega; \mathbb{C})$ and with angle 0. As a result, $\mathscr{A}$ acts from $\overset{\circ}{H}^1(\Omega; \mathbb{R})$ onto $H^{-1}(\Omega; \mathbb{R})$. This operator $\mathscr{A}$ is then regarded as a realization of the differential operator $-\sum_{j,k=1}^{n} \frac{\partial}{\partial x_k}\left[a_{jk}(x)\frac{\partial}{\partial x_j}\right]$ in $H^{-1}(\Omega; \mathbb{R})$ under the homogeneous Dirichlet boundary conditions.

In the meantime, let A be the part of $\mathscr{A}$ in $L_2(\Omega; \mathbb{C})$. Then, A is a positive definite self-adjoint operator of $L_2(\Omega; \mathbb{C})$. So, A is a positive definite self-adjoint operator of $L_2(\Omega; \mathbb{R})$ with domain $\mathscr{D}(A) \subset \overset{\circ}{H}^1(\Omega; \mathbb{R})$. This operator A is regarded as a realization of the differential operator $-\sum_{j,k=1}^{n} \frac{\partial}{\partial x_k}\left[a_{jk}(x)\frac{\partial}{\partial x_j}\right]$ in $L_2(\Omega; \mathbb{R})$ under the homogeneous Dirichlet boundary conditions.

As observed by (1.33) and (1.43), we have

$$\begin{cases} \mathscr{D}(\mathscr{A}^{\theta}) \cap H^{-1}(\Omega; \mathbb{R}) = \left([H^{-1}(\Omega; \mathbb{C}), \overset{\circ}{H}^1(\Omega; \mathbb{C})]_\theta\right)_{\mathbb{R}}, & 0 < \theta < 1, \\ \mathscr{D}(A^{\theta}) \cap L_2(\Omega; \mathbb{R}) = ([L_2(\Omega; \mathbb{C}), \mathscr{D}(A)]_\theta)_{\mathbb{R}}, & 0 < \theta < 1. \end{cases}$$

Next, consider $H^1(\Omega; \mathbb{C}) \subset L_2(\Omega; \mathbb{C})$ and let $H^1(\Omega; \mathbb{C}) \subset L_2(\Omega; \mathbb{C}) \subset H^1(\Omega; \mathbb{C})^*$ be a triplet generated by them. Consider a sesquilinear form

$$a(u, v) = \sum_{j,k=1}^{n} \int_{\Omega} a_{jk}(x) \frac{\partial u}{\partial x_j} \frac{\partial \overline{v}}{\partial x_k} dx + \int_{\Omega} c(x)u\,\overline{v}dx, \qquad u, v \in H^1(\Omega; \mathbb{C}).$$

We assume as before that $a_{jk}(x)$ satisfy (1.50)–(1.51) and $c(x)$ satisfies the condition

$$c \in L_\infty(\Omega; \mathbb{R}) \quad \text{and} \quad \text{ess.} \inf_{x \in \Omega} c(x) > 0. \tag{1.52}$$

As before, the form $a(u, v)$ satisfies (1.46)–(1.48) and all the results in the preceding subsection are available to this form $a(u, v)$.

Indeed, under (1.50)–(1.52), let $\mathscr{A}$ be the operator associated with $a(u, v)$. Then, $\mathscr{A}$ is a real sectorial operator of $H^1(\Omega; \mathbb{C})^*$ with domain $\mathscr{D}(\mathscr{A}) = H^1(\Omega; \mathbb{C})$ and with angle 0. Consequently, $\mathscr{A}$ acts from $H^1(\Omega; \mathbb{R})$ onto $H^1(\Omega; \mathbb{R})^*$. This $\mathscr{A}$ is regarded as a realization of the differential operator

$$-\sum_{j,k=1}^{n} \frac{\partial}{\partial x_k}\left[a_{jk}(x)\frac{\partial}{\partial x_j}\right] + c(x)$$

in $H^1(\Omega; \mathbb{R})^*$ under the homogeneous Neumann-type boundary conditions $\sum_{j,k=1}^{n} a_{jk}(x)v_k(x)\frac{\partial u}{\partial x_k} = 0$ on $\partial\Omega$, where $v(x) = (v_1(x), \ldots, v_n(x))$ stands for the outer normal vector at $x \in \partial\Omega$.

Meanwhile, let A be the part of $\mathscr{A}$ in $L_2(\Omega; \mathbb{C})$. Then, A is a positive definite self-adjoint operator of $L_2(\Omega; \mathbb{C})$. So, A is also a positive definite self-adjoint operator of $L_2(\Omega; \mathbb{R})$. This operator A is regarded as a realization of the differential operator $-\sum_{j,k=1}^{n} \frac{\partial}{\partial x_k}\left[a_{jk}(x)\frac{\partial}{\partial x_j}\right] + c(x)$ in $L_2(\Omega; \mathbb{R})$ under the homogeneous Neumann-type boundary conditions.

As before, by (1.33) and (1.43), we observe that

$$\begin{cases} \mathscr{D}(\mathscr{A}^\theta) \cap H^1(\Omega; \mathbb{R})^* = \left([H^1(\Omega; \mathbb{C})^*, H^1(\Omega; \mathbb{C})]_\theta\right)_{\mathbb{R}}, & 0 < \theta < 1, \\ \mathscr{D}(A^\theta) \cap L_2(\Omega; \mathbb{R}) = ([L_2(\Omega; \mathbb{C}), \mathscr{D}(A)]_\theta)_{\mathbb{R}}, & 0 < \theta < 1. \end{cases}$$

1.5 Differentiation of Operators

Let X and Y be two real Banach spaces with norm $\| \cdot \|_X$ and $\| \cdot \|_Y$ respectively. Consider an operator F from X into Y.

For $u \in X$, when there exists a bounded operator $B \in \mathscr{L}(X, Y)$ such that,

$$\text{as } h \to 0 \text{ in } X, \quad F(u + h) - F(u) - Bh = o(\|h\|_X), \tag{1.53}$$

$o(\|h\|_X)$ being a vector of Y such that, as $h \to 0$ in X, $o(\|h\|_X)/\|h\|_X \to 0$ in Y, F is said to be Fréchet differentiable at u. The operator B is called a derivative of F at u and is denoted by $F'(u)$.

Furthermore, when F is Fréchet differentiable at every vector of a neighborhood O of $u \in X$ and when the mapping $u \mapsto F'(u)$ is continuous from O into $\mathscr{L}(X, Y)$, F is said to be continuously Fréchet differentiable in O.

In the meantime, for $u \in X$, when a real variable Y-valued function

$$\theta \mapsto F(u + \theta h) \text{ is differentiable at } \theta = 0 \text{ for any } h \in X, \tag{1.54}$$

F is said to be Gâteaux differentiable at u. Put $B(u, h) = \left[\frac{d}{d\theta} F(u + \theta h)\right]_{|\theta=0} \in Y$. Then, the mapping $h \mapsto B(u, h)$ is called a derivative of F at u.

Remark 1.5 If $F : X \to Y$ is Fréchet differentiable at $u \in X$, then F is obviously Gâteaux differentiable at u with the derivative $B(u, h) = F'(u)h$. Of course, the converse is not true in general. Gâteaux derivatives $B(u, h)$ may be nonlinear or discontinuous with respect to $h \in X$. Even if $B(u, h)$ is linear and continuous in h, F is not necessarily Fréchet differentiable at u.

However, we notice the following fact.

Proposition 1.3 *Let X be a Banach space and let O be a neighborhood of a vector $\overline{u} \in X$. Consider an operator $F : O \to Y$ defined in O into a Banach space Y. If F is Gâteaux differentiable for any $u \in O$ with derivative $B(u, h)$, if $B(u, h)$ is linear and continuous in h for any $u \in O$, and if $u \mapsto B(u, \cdot)$ is continuous from O into $\mathscr{L}(X, Y)$, then F is Fréchet differentiable at $\overline{u}$.*

Proof There is a radius $r > 0$ such that the open ball $B(\overline{u}; r)$ of X with center $\overline{u}$ is such that $B(\overline{u}; r) \subset O$. Fix $h \in X$ so that $\|h\|_X < r$. Then, for any $0 \le \theta \le 1$, we see that

$$\lim_{\Delta\theta \to 0} \frac{1}{\Delta\theta} [F(\overline{u} + (\theta + \Delta\theta)h) - F(\overline{u} + \theta h)] = B(\overline{u} + \theta h, h).$$

This shows that $\theta \to F(\overline{u} + \theta h)$ is differentiable for θ with the derivative $B(\overline{u} + \theta h, h)$. By assumption, the derivative is continuous for θ. Therefore, we have

$$F(\overline{u} + h) - F(\overline{u}) = \int_0^1 B(\overline{u} + \theta h, h) d\theta.$$

Furthermore,

$$F(\overline{u} + h) - F(\overline{u}) - B(\overline{u}, h) = \int_0^1 [B(\overline{u} + \theta h, h) - B(\overline{u}, h)] d\theta.$$

Hence,

$$\|F(\overline{u} + h) - F(\overline{u}) - B(\overline{u}, h)\|_Y \le \sup_{0 \le \theta \le 1} \|B(\overline{u} + \theta h, \cdot) - B(\overline{u}, \cdot)\|_{\mathscr{L}(X,Y)} \|h\|_X.$$

This clearly shows that F is Fréchet differentiable at $\overline{u}$. $\qquad\square$

Let $\Omega \subset \mathbb{R}^n$ be a bounded domain and let $f(x, u)$ denote a real-valued function defined for $(x, u) \in \overline{\Omega} \times \mathbb{R}$. According to the definitions above, we can investigate differentiability for the operators of form $u \mapsto f(x, u)$ from $L_q(\Omega)$ into $L_p(\Omega)$, where $1 \le p \le q \le \infty$.

By $\mathscr{C}_\infty^{\overline{0,0}}(\overline{\Omega} \times \mathbb{R})$, we denote the space of real-valued functions $f(x, u)$ for $(x, u) \in \overline{\Omega} \times \mathbb{R}$ which are continuous for (x, u) and are uniformly bounded in

$\overline{\Omega} \times \mathbb{R}$ and by $\|f\|_{\mathscr{C}^{0,0}} = \sup_{(x,u) \in \overline{\Omega} \times \mathbb{R}} |f(x,u)|$ its norm. By $\mathscr{C}_\infty^{0,1}(\overline{\Omega} \times \mathbb{R})$, we denote the space of functions $f(x,u) \in \mathscr{C}_\infty^{0,0}(\overline{\Omega} \times \mathbb{R})$ which are continuously differentiable for u with partial derivative $f_u(x,u) \in \mathscr{C}_\infty^{0,0}(\overline{\Omega} \times \mathbb{R})$ and by $\|f\|_{\mathscr{C}^{0,1}} = \|f\|_{\mathscr{C}^{0,0}} + \|f_u\|_{\mathscr{C}^{0,0}}$ its norm. For $0 < \sigma < 1$, $\mathscr{C}_\infty^{0,1+\sigma}(\overline{\Omega} \times \mathbb{R})$ denotes the space of functions $f(x,u) \in \mathscr{C}_\infty^{0,1}(\overline{\Omega} \times \mathbb{R})$ whose partial derivative $f_u(x,u)$ is uniformly Hölder continuous for u with exponent σ, its norm being given by $\|f\|_{\mathscr{C}^{0,1+\sigma}} = \|f\|_{\mathscr{C}^{0,1}} + \sup_{(x,u),(x,v) \in \overline{\Omega} \times \mathbb{R}} |f_u(x,u) - f_v(x,v)|/|u-v|^\sigma$.

1.5.1 Fréchet Differentiation

We collect here some results concerning Fréchet differentiability for the operators of form $u \mapsto f(x,u)$. For the detailed proofs, however, see Section 5.1 of [Yag, Chapter 1].

Theorem 1.14 Let $f \in \mathscr{C}_\infty^{0,1+\sigma}(\overline{\Omega} \times \mathbb{R})$ $(0 < \sigma < 1)$. For $1 \le p < q < \infty$, the operator $u \mapsto f(x,u)$ from $L_q(\Omega)$ into $L_p(\Omega)$ is continuously Fréchet differentiable with the derivative $h \mapsto f_u(x,u)h$.

In this theorem, the condition that $p < q$ is essential. As seen by Theorem 1.18 below, the operator $u \mapsto f(x,u)$ from $L_p(\Omega)$ into itself is proved only to be Gâteaux differentiable (see Remark 1.5). However, we can verify the following result.

Theorem 1.15 Let $f \in \mathscr{C}^{0,1}(\overline{\Omega} \times \mathbb{R})$. The operator $u \mapsto f(x,u)$ from $\mathscr{C}(\overline{\Omega})$ into itself is continuously Fréchet differentiable with the derivative $h \mapsto f_u(x,u)h$.

As an immediate consequence of Theorem 1.14, we have the following corollary.

Corollary 1.1 Let $f \in \mathscr{C}_\infty^{0,1+\sigma}(\overline{\Omega} \times \mathbb{R})$ $(0 < \sigma < 1)$. For $q > 1$, the operator $u \mapsto \int_\Omega f(x,u)dx$ from $L_q(\Omega)$ into $\mathbb{R}$ is continuously Fréchet differentiable with the derivative $h \mapsto \int_\Omega f_u(x,u)h \, dx$.

When $n/2 < p < \infty$ and $f(u)$ is smooth, the operator $u \mapsto f(u)$ is seen to map $H_p^2(\Omega)$ into itself because of the continuous embedding $H_p^2(\Omega) \subset \mathscr{C}(\overline{\Omega})$. Then, we have the following theorems.

Theorem 1.16 Let $f \in \mathscr{C}^3(\mathbb{R})$. For $n/2 < p < \infty$, the operator $u \mapsto f(u)$ from $H_p^2(\Omega)$ into itself is continuously Fréchet differentiable with the derivative $h \mapsto f'(u)h$.

Theorem 1.17 Let $f \in \mathscr{C}^{m+1}(\mathbb{R})$ with some integer m. For $1 \le s$, $n/2 < s \le m$, the operator $u \mapsto f(u)$ from $H^s(\Omega)$ into itself is continuously Fréchet differentiable with the derivative $h \mapsto f'(u)h$.

1.5.2 Gâteaux Differentiation

Let us now see Gâteaux differentiability for the operator $u \mapsto f(x, u)$, where $f(x, u) \in \mathscr{C}^{0,1}_{\infty}(\overline{\Omega} \times \mathbb{R})$, from $L_p(\Omega)$ into itself.

Theorem 1.18 *Let* $f \in \mathscr{C}^{0,1}_{\infty}(\overline{\Omega} \times \mathbb{R})$. *For* $1 \leq p < \infty$, *the operator* $u \mapsto f(x, u)$ *from* $L_p(\Omega)$ *into itself is Gâteaux differentiable with the derivative* $B(u, h) = f_u(x, u)h$. *The derivative is a bounded linear operator of* $L_p(\Omega)$ *and is strongly continuous, that is, for each fixed* $h \in L_p(\Omega)$, *as* $u \to \overline{u}$ *in* $L_p(\Omega)$, $B(u, h) \to B(\overline{u}, h)$ *in* $L_p(\Omega)$.

For the proof, see [Yag, Theorem 1.18].

As an immediate consequence, we verify the following corollaries.

Corollary 1.2 *Let* $f \in \mathscr{C}^{0,1}_{\infty}(\overline{\Omega} \times \mathbb{R})$. *For* $1 \leq p < \infty$, *if* $u \in \mathscr{C}^1([0, T]; L_p(\Omega))$, *then* $f(x, u(t)) \in \mathscr{C}^1([0, T]; L_p(\Omega))$ *with* $\frac{d}{dt} f(x, u(t)) = f_u(x, u(t))u'(t)$.

Corollary 1.3 *Let* $f \in \mathscr{C}^{0,1}_{\infty}(\overline{\Omega} \times \mathbb{R})$. *If* $u \in \mathscr{C}^1([0, T]; L_1(\Omega))$, *then it holds that* $\int_{\Omega} f(x, u(t))dx \in \mathscr{C}^1([0, T])$ *with* $\frac{d}{dt} \int_{\Omega} f(x, u(t))dx = \int_{\Omega} f_u(x, u(t))u'(t)dx$.

1.6 Some Other Materials

We will collect some other materials we need in this monograph.

1.6.1 Fredholm Operators

Let X and Y be two real Banach spaces. A bounded linear operator T from X into Y is called *a Fredholm operator* if its kernel $\mathscr{K}(T)$ is a subspace of X of finite dimension and its range $\mathscr{R}(T)$ is a closed subspace of Y of finite codimension. The following theorem in the Riesz–Schauder theory gives an important class of Fredholm operators.

Theorem 1.19 *Let* K *be a compact linear operator from* X *into itself. Then,* $T = I - K$ *is a Fredholm operator, where* I *denotes the identity operator of* X. *Moreover, it holds that* $\dim \mathscr{K}(T) = \operatorname{codim} \mathscr{R}(T)$.

For the proof of theorem, see [DL84a, Lemme II.4.3] or [Bre11, Theorem 6.6].

As an immediate consequence of this theorem, we have the following corollary.

Corollary 1.4 *Let a bounded linear operator* T *from* X *into* Y *be given as* $T = S - K$, *where* S *is an isomorphism from* X *onto* Y *and* K *is a compact operator from* X *into* Y. *Then,* T *is a Fredholm operator with* $\dim \mathscr{K}(T) = \operatorname{codim} \mathscr{R}(T)$.

1.6.2   Łojasiewicz Gradient Inequality

Let Ω denote a domain of $\mathbb{R}^N$. Consider a real-valued smooth function $\phi(\xi)$ for $\xi = (\xi_1, \ldots, \xi_N) \in \Omega$ and let $\bar{\xi} \in \Omega$ be its critical point, i.e., $\nabla \phi(\bar{\xi}) = 0$. So, the surface $\xi_{N+1} = \phi(\xi)$ in the space $\mathbb{R}^N \times \mathbb{R}$ is tangential to the plane $\xi_{N+1} = \phi(\bar{\xi})$ at $\bar{\xi}$. When $\phi(\xi)$ is analytic in Ω, the norm of $\nabla \phi(\xi)$ can be estimated from below by a power function of $|\phi(\xi) - \phi(\bar{\xi})|$ in a neighborhood of $\bar{\xi}$.

Theorem 1.20 *Let $\phi : \Omega \to \mathbb{R}$ be an analytic function defined in a domain $\Omega \subset \mathbb{R}^N$ and let $\bar{\xi} \in \Omega$ be its critical point. Then, there exist an exponent $0 < \theta \le \frac{1}{2}$ and a neighborhood $U \subset \Omega$ of $\bar{\xi}$ such that the inequality*

$$\|\nabla \phi(\xi)\|_{\mathbb{R}^N} \ge D|\phi(\xi) - \phi(\bar{\xi})|^{1-\theta}, \qquad \xi \in U, \tag{1.55}$$

holds true with some constant $D > 0$.

This inequality was first presented by Łojasiewicz [Loj63, Loj65]. So, (1.55) is called *the Łojasiewicz gradient inequality*.

1.7   Notes

The basic materials of Complex Functional Analysis presented in Sect. 1.1 were all explained or obtained in [Yag10]. As for dual spaces and adjoint spaces, see Subsections 6.1 and 6.2 of [Yag10, Chapter 1] respectively. Interpolation spaces are explained in Subsection 5.1 of [Yag10, Chapter 1], see also [Tri78, Section 1.9]. Triplets of spaces are constructed in Subsection 7.1 of [Yag10, Chapter 1]. Sectorial operators are defined and studied in [Yag10, Chapter 2]. Theorem 1.4 is verified by Theorems 4.1 and 4.4 in [Yag10, Chapter 4].

The notion of conjugations in complex Banach spaces and the notion of real operators acting in conjugated Banach spaces were originally presented by the paper [Yag17]. These various properties explained in Sects. 1.2 and 1.3 were also obtained there. As those are relatively new and give us important frameworks of study, we presented them again here, together with their full proofs.

Let T be a compact operator of a Banach space. By virtue of the Riesz–Schauder theory, we can know the structure of its spectrum $\sigma(T)$ or the necessary and sufficient condition for solvability of the vector equation $Tu = f$, f being a given vector and u being an unknown vector, in terms of its dual operator T'. Theorem 1.19 is then shown in that theory and plays an important role.

Except for the case where $N = 1$, the proof of the gradient inequality (1.55) is not at all elementary. As for its proof, see also Łojasiewicz–Zurro [LZ99].

Chapter 2
Asymptotic Convergence

In this chapter, we will introduce our general settings and show the main convergence theorem. But, before doing so, it might be better to study a semilinear heat equation as a typical equation for understanding the essence of our arguments.

2.1 Semilinear Heat Equation

Consider a three-dimensional semilinear heat equation

$$\begin{cases} \dfrac{\partial u}{\partial t} - a\Delta u = f(u) & \text{in} \quad \Omega \times (0, \infty), \\ u = 0 & \text{on} \quad \partial\Omega \times (0, \infty), \end{cases} \tag{2.1}$$

in a convex or $\mathscr{C}^2$, bounded domain $\Omega \subset \mathbb{R}^3$. Here, $f(u)$ is a smooth function defined for $-\infty < u < \infty$ which has at least two zeros $u_- < u_+$ such that $-\infty < u_- \leq 0 \leq u_+ < \infty$. And $u = u(x, t)$ is an unknown function to which the homogeneous Dirichlet conditions are imposed on the boundary $\partial\Omega$ of Ω. Imposing also an initial condition

$$u(x, 0) = u_0(x) \qquad \text{in} \quad \Omega, \tag{2.2}$$

we are concerned with the initial-boundary-value problem (2.1)–(2.2).

Let us apply the functional analytic methods to the problem (2.1)–(2.2). Equation (2.1) is formulated as an evolution equation

$$u' + Au = f(u), \qquad 0 < t < \infty, \tag{2.3}$$

© The Author(s), under exclusive license to Springer Nature Singapore Pte Ltd. 2021
A. Yagi, *Abstract Parabolic Evolution Equations and Łojasiewicz-Simon Inequality I*,
SpringerBriefs in Mathematics, https://doi.org/10.1007/978-981-16-1896-3_2

in the space $L_2(\Omega)$. Here, A denotes a realization of $-a\Delta$ in $L_2(\Omega)$ under the boundary conditions $u = 0$ on $\partial\Omega$ (see Sect. 1.4), $A \geq c > 0$ being a positive definite self-adjoint operator of $L_2(\Omega)$. Its domain is characterized by $\mathscr{D}(A) = H^2(\Omega) \cap \overset{\circ}{H}{}^1(\Omega)$, and $\mathscr{D}(A)$ is equipped with the graph norm $\|Au\|_{L_2} = \|a\Delta u\|_{L_2}$, $u \in \mathscr{D}(A)$. The square root $A^{\frac{1}{2}}$ of A has a domain $\mathscr{D}(A^{\frac{1}{2}}) = \overset{\circ}{H}{}^1(\Omega)$, and its graph norm is given by $\|A^{\frac{1}{2}}u\|_{L_2} = \sqrt{a}\,\|\nabla u\|_{L_2}$, $u \in \mathscr{D}(A^{\frac{1}{2}})$, because it holds for $u \in \mathscr{D}(A)$ that $\|A^{\frac{1}{2}}u\|^2_{L_2} = (Au, u)_{L_2} = (-a\Delta u, u)_{L_2} = a\|\nabla u\|^2_{L_2}$. Meanwhile, $f : u \mapsto f(u)$ is a nonlinear operator of $L_2(\Omega)$ with domain $\mathscr{D}(f) = \{u \in L_2(\Omega);\ f(u) \in L_2(\Omega)\}$.

For constructing a local solution to (2.3), it is possible to apply Theorem 1.4. Consider (1.17) in the space $L_2(\Omega; \mathbb{C})$ in which A is the realization of $-a\Delta$ in $L_2(\Omega; \mathbb{C})$ under the boundary conditions $u = 0$ on $\partial\Omega$ and f is a nonlinear operator $u \mapsto f(\mathrm{Re}\,u)$ of $L_2(\Omega; \mathbb{C})$ with domain $\mathscr{D}(f) = \{u \in L_2(\Omega; \mathbb{C});\ f(\mathrm{Re}\,u) \in L_2(\Omega)\}$. Fix an exponent $\frac{3}{4} < \beta < 1$ so that $\mathscr{D}(A^\beta) \subset H^{2\beta}(\Omega; \mathbb{C}) \subset \mathscr{C}(\overline{\Omega}; \mathbb{C})$; then, the condition (1.18) is satisfied with $\beta = \eta$. Take an initial value u_0 from $\mathscr{D}(A^\beta)$. Then, there exists a unique local solution u to (1.17). But, as A is a real operator, if $u_0 \in \mathscr{D}(A^\beta)$ is real, then the local solution u must also be real. By (1.19), this u is a unique local solution to (2.2)–(2.3) in the function space

$$u \in \mathscr{C}((0, T_{u_0}]; \mathscr{D}(A)) \cap \mathscr{C}([0, T_{u_0}]; \mathscr{D}(A^\beta)) \cap \mathscr{C}^1((0, T_{u_0}]; L_2(\Omega)). \qquad (2.4)$$

By (1.20), the solution satisfies also the estimate

$$t^{1-\beta}\|Au(t)\|_{L_2} + \|A^\beta u(t)\|_{L_2} \leq C_{u_0}, \qquad 0 < t \leq T_{u_0}. \qquad (2.5)$$

Here, T_{u_0} and C_{u_0} are determined by the magnitude of $\|A^\beta u_0\|_{L_2}$ alone.

When an initial function u_0 be such that

$$u_- \leq u_0(x) \leq u_+ \qquad \text{in } \Omega, \qquad (2.6)$$

we can verify the following comparison theorem.

Proposition 2.1 *Let $u_0 \in \mathscr{D}(A^\beta)$ satisfy (2.6). Let u be any local solution to (2.2)–(2.3) given on $[0, T_u]$ on which u lies in the space (2.4) (but T_u being substituted for T_{u_0}). Then, actually $u(t)$ must satisfy the same condition as (2.6) for every $0 < t \leq T_u$.*

Proof In the proof, we will utilize a $\mathscr{C}^1$ cutoff function $H(\xi)$ defined by

$$H(\xi) = \tfrac{1}{2}\xi^2 \quad \text{for} \quad -\infty < \xi < 0 \quad \text{and} \quad H(\xi) = 0 \quad \text{for} \quad 0 \leq \xi < \infty.$$

Consider the function

$$\varphi(t) = \int_\Omega H(u(t) - u_-)dx, \qquad 0 \leq t \leq T_u.$$

Then, $\varphi(t)$ is continuously differentiable for $0 < t \leq T$ with the derivative

$$\varphi'(t) = \int_\Omega H'(u(t) - u_-)u'(t)dx = \int_\Omega H'(u(t) - u_-)[a\Delta u(t) + f(u(t))]dx.$$

Here, we have

$$a\int_\Omega H'(u(t) - u_-)\Delta u(t)dx = -a\int_\Omega \nabla H'(u(t) - u_-) \cdot \nabla u(t)dx$$

$$+ a\int_{\partial\Omega} H'(u(t) - u_-)\tfrac{\partial}{\partial n}u(t)dx.$$

Since $H'(u(t) - u_-) = H'(-u_-) = 0$ on $\partial\Omega$ and $\nabla H'(u(t) - u_-) = H''(u(t) - u_-)\nabla u(t)$ in Ω (see [Yag10, (1.96)]), it follows that $a\int_\Omega H'(u(t) - u_-)\Delta u(t)dx \leq 0$. Meanwhile,

$$\int_\Omega H'(u(t) - u_-)f(u(t))dx = \int_\Omega H'(u(t) - u_-)[f(u(t)) - f(u_-)]dx$$

$$\leq C_u \int_\Omega |H'(u(t) - u_-)[u(t) - u_-]|dx$$

with some constant $C_u > 0$. But, because of $H'(\xi)\xi = 2H(\xi)$ for $-\infty < \xi < \infty$, it follows that $\int_\Omega H'(u(t) - u_-)f(u(t))dx \leq 2C_u\varphi(t)$.

We therefore observe that $\varphi'(t) \leq 2C_u\varphi(t)$; consequently, $\varphi(t) \leq e^{2C_u t}\varphi(0)$ for $0 \leq t \leq T_u$. Since $\varphi(0) = 0$ due to (2.6), it is concluded that $\varphi(t) = 0$, namely, $u_- \leq u(t)$ for every $0 < t \leq T_u$.

We can argue in a quite analogous way for the function $\psi(t) = \int_\Omega H(u_+ - u(t))dx$, $0 \leq t \leq T_u$, to conclude that $u(t) \leq u_+$ for every $0 < t \leq T_u$. □

Furthermore, we can obtain an *a priori* estimate for the local solutions.

Proposition 2.2 *Let $u_0 \in \mathscr{D}(A^\beta)$ satisfy (2.6). Let u be any local solution to (2.2)–(2.3) given on an interval $[0, T_u]$ on which u lies in the space (2.4) (but T_u being substituted for T_{u_0}). Then, $u(t)$ must satisfy the estimate*

$$t^{1-\beta}\|Au(t)\|_{L_2} + \|A^\beta u(t)\|_{L_2} \leq C_{u_0}, \qquad 0 < t \leq T_u, \tag{2.7}$$

with a constant C_{u_0} independent of T_u.

Proof We use the representation formula

$$u(t) = e^{-tA}u_0 + \int_0^t e^{-(t-s)A} f(u(s))ds, \qquad 0 \leq t \leq T_u,$$

in $L_2(\Omega)$, where e^{-tA} is the analytic semigroup generated by $-A$ (see (1.45)). It then follows that

$$\|A^\beta u(t)\|_{L_2} \leq \|e^{-tA}A^\beta u_0\|_{L_2} + \int_0^t \|A^\beta e^{-(t-s)A} f(u(s))\|_{L_2} ds.$$

Since $\|f(u(s))\|_{L_2} \leq |\Omega|^{\frac{1}{2}} \max_{u_- \leq \xi \leq u_+} |f(\xi)|$, $\|A^\beta u(t)\|_{L_2}$ is estimated by

$$\|A^\beta u(t)\|_{L_2} \leq e^{-ct}\|A^\beta u_0\|_{L_2} + |\Omega|^{\frac{1}{2}} \max_{u_- \leq \xi \leq u_+} |f(\xi)| \int_0^t (t-s)^{-\beta} e^{-c(t-s)} ds.$$

Because of $\int_0^t (t-s)^{-\beta} e^{-c(t-s)} ds = \int_0^t s^{-\beta} e^{-cs} ds < \int_0^\infty s^{-\beta} e^{-cs} ds < \infty$, the second estimate of (2.7) is proved.

Meanwhile, the first estimate is obtained by applying (2.5) repeatedly. Let $0 < \tau < T_u$ be any time; then, as u can be regarded as a local solution to (2.3) on the interval $[\tau, \tau + T_{u(\tau)}]$, (2.5) can yield the estimate $\|Au(t)\| \leq C_{u(\tau)}(t - \tau)^{\beta-1}$ for $\tau < t \leq \tau + T_{u(\tau)}$, where $T_{u(\tau)}$ and $C_{u(\tau)}$ depend only on $\|A^\beta u(\tau)\|_{L_2}$, namely, only on $\|A^\beta u_0\|_{L_2}$. Hence, the first estimate of (2.7) is also proved. □

Let $u_0 \in \mathscr{D}(A^\beta)$ satisfy (2.6). As the estimate (2.7) provides an *a priori* estimate for the local solutions, the standard arguments conclude that there exists a unique global solution to (2.2)–(2.3) in the function space

$$u \in \mathscr{C}((0, \infty); \mathscr{D}(A)) \cap \mathscr{C}([0, \infty); \mathscr{D}(A^\beta)) \cap \mathscr{C}^1((0, \infty); L_2(\Omega)), \qquad (2.8)$$

and the solution u satisfies the estimate

$$\min\{t^{1-\beta}, 1\}\|Au(t)\|_{L_2} + \|A^\beta u(t)\|_{L_2} \leq C_{u_0}, \qquad 0 < t < \infty. \qquad (2.9)$$

In what follows, we fix an initial function $u_0(x)$ taken from $\mathscr{D}(A^\beta)$ and satisfying (2.6). Of course, Proposition 2.1 is available to the global solution $u(t)$, namely, it holds that $u_- \leq u(t) \leq u_+$ for every $0 < t < \infty$. Then, the values of $f(u)$ for $u \notin [u_-, u_+]$ make no contribution to the asymptotic behavior of $u(t)$. It is allowed to cut off the values $f(u)$ outside the interval $[u_-, u_+]$. For this reason, we will assume that $f(u)$ vanishes identically for $u \in (-\infty, u_- - 1] \cup [u_+ + 1, \infty)$.

There exists a Lyapunov function for the global solution. Take the inner product of $L_2(\Omega)$ for the equation of (2.3) and $u'(t)$. By Corollary 1.3, it follows from $\|u'(t)\|_{L_2}^2 + (Au(t), u'(t))_{L_2} = (f(u(t)), u'(t))_{L_2}$ that

$$\frac{1}{2}\frac{d}{dt}\|A^{\frac{1}{2}}u(t)\|_{L_2}^2 - \frac{d}{dt}\int_\Omega F(u(t))dx = -\|u'(t)\|_{L_2}^2, \qquad 0 < t < \infty,$$

where $F(u) = \int_0^u f(v)dv$, $-\infty < u < \infty$. This shows that the function

$$\Phi(u) = \int_\Omega \left[\frac{a}{2}|\nabla u|^2 - F(u)\right]dx, \qquad u \in \mathscr{D}(A^{\frac{1}{2}}),$$

is a Lyapunov function for $u(t)$. For the mapping $u \mapsto \int_\Omega F(u)dx$, Corollary 1.1 is available with $q = 2$ to observe that $\int_\Omega[F(u+h) - F(u)]dx = \int_\Omega f(u)h\,dx + o(\|h\|_{L_2})$ for $u, h \in L_2(\Omega)$. It is then possible to obtain that

$$\Phi(u+h) - \Phi(u) - (A^{\frac{1}{2}}[u - A^{-1}f(u)], A^{\frac{1}{2}}h)_{L_2} = o(\|h\|_{L_2}), \qquad u, h \in \mathscr{D}(A^{\frac{1}{2}}).$$

Here, we want to consider the three spaces $\mathscr{D}(A) \subset \mathscr{D}(A^{\frac{1}{2}}) \subset L_2(\Omega)$ as a triplet of real spaces defined by Definition 1.2. (Now, $\{\mathscr{D}(A), L_2(\Omega)\}$ is seen to be an adjoint pair with the duality product $\langle u, f\rangle_{\mathscr{D}(A), L_2} = (Au, f)_{L_2}$.)

The formula observed above then shows that $\Phi : \mathscr{D}(A^{\frac{1}{2}}) \to \mathbb{R}$ is continuously differentiable in the sense that, as $h \to 0$ in $\mathscr{D}(A^{\frac{1}{2}})$,

$$\Phi(u+h) - \Phi(u) - ([u - A^{-1}f(u)], h)_{\mathscr{D}(A^{\frac{1}{2}})} = o(\|h\|_{\mathscr{D}(A^{\frac{1}{2}})}),$$

which means that $\Phi'(u) = J[u - A^{-1}f(u)]$, where $J : \mathscr{D}(A^{\frac{1}{2}}) \to \mathscr{D}(A^{\frac{1}{2}})'$ is an isometric isomorphism defined by (1.38). So, let us put $\dot{\Phi}(u) = J^{-1}\Phi'(u)$ for $u \in \mathscr{D}(A^{\frac{1}{2}})$. Then, it is clear that $u \in \mathscr{D}(A)$ implies $\dot{\Phi}(u) \in \mathscr{D}(A)$ together with $A\dot{\Phi}(u) = Au - f(u)$. In view of (2.4), we hence arrive at the equalities

$$-\frac{d}{dt}\Phi(u(t)) = \|u'(t)\|_{L_2}^2 = \|A\dot{\Phi}(u(t))\|_{L_2}^2, \qquad 0 < t < \infty. \tag{2.10}$$

The ω-limit set of the global solution $u(t)$ is defined as usual by

$$\omega(u) = \{\overline{u}; \exists t_n \nearrow \infty \text{ such that } u(t_n) \to \overline{u} \text{ in } L_2(\Omega)\}. \tag{2.11}$$

As seen by (2.5), the trajectory of $u(t)$ $(1 \le t < \infty)$ is contained in a closed ball $\overline{B}^{\mathscr{D}(A)}(0; C_{u_0})$ which is compact in $L_2(\Omega)$ and weakly sequentially compact in $\mathscr{D}(A)$. Hence, it is verified that $\omega(u) \ne \emptyset$ and $\omega(u) \subset \overline{B}^{\mathscr{D}(A)}(0; C_{u_0})$ and that, if $\overline{u} \in \omega(u)$, then $u(t_n) \to \overline{u}$ in $\mathscr{D}(A^\theta)$ for any $0 < \theta < 1$. In particular, it follows that $\Phi(u(t_n)) \to \Phi(\overline{u})$. We thus observe that

$$\lim_{t \to \infty} \Phi(u(t)) = \inf_{0 \le t < \infty} \Phi(u(t)) = \Phi(\overline{u}), \qquad \forall \overline{u} \in \omega(u). \tag{2.12}$$

In view of this, integrate the equality (2.10) for $0 \leq t < \infty$. Then, we have

$$\int_0^\infty \|u'(t)\|_{L_2}^2 dt = \Phi(u_0) - \inf_{0 \leq t < \infty} \Phi(u(t)) < \infty. \tag{2.13}$$

So, for any number $\varepsilon > 0$ and any time $T > 0$, there must exist a time $\tau > T$ such that $\|u'(\tau)\|_{L_2}^2 < \varepsilon$; as a consequence, there exists a temporal sequence $t_n \nearrow \infty$ such that $u'(t_n) \to 0$ in $L_2(\Omega)$ and that $u(t_n) \to \overline{u}$ for some $\overline{u} \in \omega(u)$. As $Au(t_n) - f(u(t_n)) \to A\overline{u} - f(\overline{u})$ weakly in $L_2(\Omega)$, this ω-limit must satisfy $A\overline{u} - f(\overline{u}) = 0$, that is, $\overline{u}$ is a stationary solution of (2.3). Furthermore, as $\Phi'(\overline{u}) = 0$, $\overline{u}$ is also a critical point of the function Φ.

In what follows, we fix this ω-limit $\overline{u}$ and want to prove that, as a matter of fact, $\omega(u) = \{\overline{u}\}$, i.e., $\omega(u)$ is a singleton. If $\frac{d}{dt}\Phi(u(t)) = 0$ at some $t = \overline{t}$, then (2.10) implies that $u(\overline{t})$ is a stationary solution of (2.3) and therefore $\omega(u) = \{u(\overline{t})\}$ is a singleton. So, let us show the result under a condition that $\frac{d}{dt}\Phi(u(t)) < 0$ for any $0 < t < \infty$. Then, in view of (2.12), we have $\Phi(u(t)) > \Phi(\overline{u})$ for any $0 < t < \infty$.

Consider a function $[\Phi(u(t)) - \Phi(\overline{u})]^\theta$ for $0 < t < \infty$, where $0 < \theta < 1$ is an exponent to be fixed below. In view of (2.10), we have

$$-\frac{d}{dt}[\Phi(u(t)) - \Phi(\overline{u})]^\theta = -\theta[\Phi(u(t)) - \Phi(\overline{u})]^{\theta-1}\frac{d}{dt}\Phi(u(t))$$

$$= \theta[\Phi(u(t)) - \Phi(\overline{u})]^{\theta-1}\|u'(t)\|_{L_2}\|A\dot{\Phi}(u(t))\|_{L_2}.$$

Here, let us assume that a fundamental gradient inequality of Φ of the form

$$\|A\dot{\Phi}(u(t))\|_{L_2} \geq D|\Phi(u(t)) - \Phi(\overline{u})|^{1-\theta} \quad \text{if} \quad u(t) \in B^{\mathscr{D}(A^{\frac{1}{2}})}(\overline{u}; r), \tag{2.14}$$

holds true with some exponent $0 < \theta \leq \frac{1}{2}$ in a ball $B^{\mathscr{D}(A^{\frac{1}{2}})}(\overline{u}, r)$ with center $\overline{u}$ and radius $r > 0$, D being a positive constant. Under (2.14), if $u(t)$ is sufficiently close to $\overline{u}$, then it follows that

$$-\frac{d}{dt}[\Phi(u(t)) - \Phi(\overline{u})]^\theta \geq D\theta\|u'(t)\|_{L_2}.$$

Assume that $0 < s < t$ are such that $u(\tau) \in B^{\mathscr{D}(A^{\frac{1}{2}})}(\overline{u}; r)$ for all $\tau \in [s, t]$. Then,

$$[\Phi(u(s)) - \Phi(\overline{u})]^\theta - [\Phi(u(t)) - \Phi(\overline{u})]^\theta \geq D\theta\int_s^t \|u'(\tau)\|_{L_2}d\tau.$$

Consequently,

$$\|u(t) - u(s)\|_{L_2} \leq (D\theta)^{-1}[\Phi(u(s)) - \Phi(\overline{u})]^\theta. \tag{2.15}$$

This estimate in fact provides an important result for the trajectory of $u(t)$. In view of (2.11) and (2.12), take a time $t_N \geq 1$ such that $\|A^{\frac{1}{2}}[u(t_N) - \overline{u}]\|_{L_2} \leq \frac{r}{3}$ and that $(2C_{u_0}D^{-1}\theta^{-1})^{\frac{1}{2}}[\Phi(u(t_N)) - \Phi(\overline{u})]^{\frac{\theta}{2}} \leq \frac{r}{3}$. Then, (2.15) (with $s = t_N$) implies that, if $u(t)$ is such that $u(\tau) \in B^{\mathscr{D}(A^{\frac{1}{2}})}(\overline{u}; r)$ for all $\tau \in [t_N, t]$, then

$$\|A^{\frac{1}{2}}[u(t) - u(t_N)]\|_{L_2} \leq \|A[u(t) - u(t_N)]\|_{L_2}^{\frac{1}{2}}\|u(t) - u(t_N)\|_{L_2}^{\frac{1}{2}} \leq \frac{r}{3}.$$

Hence, $\|A^{\frac{1}{2}}[u(t) - \overline{u}]\|_{L_2} \leq \|A^{\frac{1}{2}}[u(t) - u(t_N)]\|_{L_2} + \|A^{\frac{1}{2}}[u(t_N) - \overline{u}]\|_{L_2} \leq \frac{2r}{3}$, which means that after the time t_N, the trajectory of $u(t)$ is trapped in $B^{\mathscr{D}(A^{\frac{1}{2}})}(\overline{u}; r)$.

Now, let $t_N \leq s < t_n$ and use again (2.15) (with $t = t_n$). Then, after letting $t_n \nearrow \infty$, we conclude that

$$\|\overline{u} - u(s)\|_{L_2} \leq (D\theta)^{-1}[\Phi(u(s)) - \Phi(\overline{u})]^{\theta}, \qquad \forall s \geq t_N.$$

We have thus proved that, as $t \to \infty$, the global solution $u(t)$ tends to a stationary $\overline{u}$ of (2.3), although the fundamental gradient inequality (2.14) remains to be verified. As will be shown in Chap. 3, the inequality which is called *the Łojasiewicz–Simon Inequality* is generally derived from twice continuous Fréchet differentiability of Φ and analyticity of Φ on its critical manifold. In the present case, the additional assumption that

$$f(u) \text{ is analytic in some neighborhood of the interval } [u_-, u_+] \qquad (2.16)$$

can indeed provide the inequality. The detailed derivation of (2.14) from (2.16) will be described in Section 3.6 of [Yag].

2.2 General Settings

Taking account of the study on Eq. (2.1), let us set up the general frameworks of our arguments.

Let X be a real Hilbert space with inner product $(\cdot, \cdot)_X$ and norm $\|\cdot\|_X$, and let Z be a second real Hilbert space with norm $\|\cdot\|_Z$, Z being embedded in X densely and continuously. According to Definition 1.2 (see also Remark 1.4), we introduce the triplet

$$Z \subset X \subset Z^* \qquad (2.17)$$

of real spaces. The norm of Z^* is denoted by $\|\cdot\|_{Z^*}$ and the duality product between Z and Z^* is denoted by $\langle\cdot,\cdot\rangle_{Z\times Z^*}$. Then, $(u, f)_X = \langle u, f\rangle_{Z\times Z^*}$ holds for $u \in Z$ and $f \in X$ due to (1.9). By the property (4) of Theorem 1.10, we have

$$\|u\|_X \leq C\|u\|_Z^{\frac{1}{2}}\|u\|_{Z^*}^{\frac{1}{2}}, \qquad u \in Z, \tag{2.18}$$

with some constant $C > 0$.

Let $u(t)$ be a function defined on $[0, \infty)$ which lies in the function space

$$u \in \mathscr{C}([0, \infty); Z) \cap \mathscr{C}^1([0, \infty); Z^*). \tag{2.19}$$

Assume that $u(t)$ satisfies a global norm estimate

$$\|u(t)\|_Z \leq R \quad \text{for all} \quad 0 \leq t < \infty, \tag{2.20}$$

with some constant $R > 0$.

Meanwhile, let us introduce a functional $\Phi(f)$ defined for $f \in X$ with values in $\mathbb{R}$. Assume that this functional is continuously differentiable in the sense that, for each $f \in X$, there exists a vector $\dot\Phi(f) \in X$ for which it holds that

$$\text{as } h \to 0 \text{ in } X, \quad \Phi(f + h) - \Phi(f) - (\dot\Phi(f), h)_X = o(\|h\|_X), \tag{2.21}$$

and that

$$\text{the mapping } f \mapsto \dot\Phi(f) \text{ is continuous from } X \text{ into itself.} \tag{2.22}$$

As a matter of fact, these (2.21)–(2.22) are equivalent to the condition:

$$\Phi : X \to \mathbb{R} \text{ is continuously Fréchet differentiable.} \tag{2.23}$$

Indeed, recall the isometric isomorphism $J : X \to X'$ presented in (1.38) and put $\dot\Phi(f) = J^{-1}\Phi'(f)$ for $f \in X$. Then, by (1.2), we can write as $[\Phi'(f)](h) = (h, J^{-1}\Phi'(f))_X = (\dot\Phi(f), h)_X$. Hence, by definition (1.53), (2.23) certainly yields (2.21)–(2.22). The converse is also obvious.

Furthermore, assume that, when $u \in Z$, the derivative $\dot\Phi(u)$ is in Z and that

$$\text{the mapping } u \mapsto \dot\Phi(u) \text{ is continuous from } Z \text{ into itself.} \tag{2.24}$$

Similarly, this is implied by the conditions that, when $u \in Z$, the derivative $\Phi'(u)$ is in $(Z^*)' \subset X'$ and that

$$\text{the mapping } u \mapsto \Phi'(u) \text{ is continuous from } Z \text{ into } (Z^*)'. \tag{2.25}$$

Indeed, use again the operator J in (1.38) which is also an isometric isomorphism from Z onto $(Z^*)'$.

Proposition 2.3 *Under the situations above,* $\Phi(u(t))$ *is continuously differentiable for* $0 \le t < \infty$ *and the derivative is given by*

$$\frac{d}{dt}\Phi(u(t)) = \left\langle \dot{\Phi}(u(t)), u'(t) \right\rangle_{Z \times Z^*}, \qquad 0 \le t < \infty. \tag{2.26}$$

Proof The function $\theta \mapsto \Phi(\theta u(t) + (1 - \theta)u(s))$ is continuously differentiable for $0 \le \theta \le 1$ and it holds by (2.21) that

$$\Phi(u(t)) - \Phi(u(s)) = \int_0^1 (\dot{\Phi}(\theta u(t) + (1 - \theta)u(s)), u(t) - u(s))_X d\theta.$$

Using the duality product, we have

$$\frac{\Phi(u(t)) - \Phi(u(s))}{t - s} = \left\langle \dot{\Phi}(u(s)), \frac{u(t) - u(s)}{t - s} \right\rangle_{Z \times Z^*}$$

$$+ \int_0^1 \left\langle \dot{\Phi}((\theta u(t) + (1 - \theta)u(s)) - \dot{\Phi}(u(s)), \frac{u(t) - u(s)}{t - s} \right\rangle_{Z \times Z^*} d\theta.$$

Here let $t \to s$. Then, (2.19) and (2.24) yield the desired equality (2.26). $\qquad \square$

As usual, the ω-limit set of the function $u(t)$ is defined by

$$\omega(u) = \{\overline{u} \in Z^*; \ \exists t_n \nearrow \infty \ \text{such that} \ u(t_n) \to \overline{u} \ \text{in} \ Z^*\}. \tag{2.27}$$

As the closed ball $\overline{B}^Z(0; R)$ is sequentially, weakly compact in Z, we can assume that, if $\overline{u} \in \omega(u)$, then $u(t_n) \to \overline{u}$ weakly in Z; in particular, $\overline{u} \in \overline{B}^Z(0; R)$. Furthermore, by (2.18) and (2.20), $u(t_n) \to \overline{u}$ in Z^* also implies $u(t_n) \to \overline{u}$ in X. Therefore,

$$\lim_{n \to \infty} \Phi(u(t_n)) = \Phi(\overline{u}). \tag{2.28}$$

2.3 Structural Assumptions

Taking account of the study on Eq. (2.1) again, let us make the structural assumptions of our methods concerning the function $u(t)$ and the Lyapunov function $\Phi(u)$.

We will assume that there exists an ω-limit $\overline{u}$ of $u(t)$ satisfying the following four conditions:

(I) *Critical Condition.* The $\overline{u}$ is a critical point of $\Phi(u)$, i.e., $\Phi'(\overline{u}) = 0$ (which is equivalent to $\dot{\Phi}(\overline{u}) = 0$).

(II) *Lyapunov Function.* There exists a radius $r' > 0$ such that

$$\Phi(u(t)) > \Phi(\overline{u}) \quad \text{and} \quad \frac{d}{dt}\Phi(u(t)) \leq 0 \quad \text{if } u(t) \in B^X(\overline{u}; r'). \quad (2.29)$$

(III) *Angle Condition.* There exist a radius $r'' > 0$ and a constant $\delta > 0$ such that

$$-\langle \dot{\Phi}(u(t)), u'(t) \rangle_{Z \times Z^*} \geq \delta \|\dot{\Phi}(u(t))\|_Z \|u'(t)\|_{Z^*} \quad \text{if } u(t) \in B^X(\overline{u}; r''). \tag{2.30}$$

(IV) *Gradient Inequality.* There exist a radius $r''' > 0$ and an exponent $0 < \theta \leq \frac{1}{2}$ for which it holds that

$$\|\dot{\Phi}(u(t))\|_Z \geq D|\Phi(u(t)) - \Phi(\overline{u})|^{1-\theta} \quad \text{if } u(t) \in B^X(\overline{u}; r'''), \quad (2.31)$$

$D > 0$ being some constant.

Under these assumptions, put $r = \min\{r', r'', r'''\}$. Then, when the trajectory $u(t)$ lies in the ball $B^X(\overline{u}; r)$, the difference $\|u(t) - u(s)\|_{Z^*}$ for $t > s$ can be controlled by the values of the Lyapunov function $\Phi(u(s)) - \Phi(\overline{u})$. In fact, we have the following proposition.

Proposition 2.4 *For $s < t$, let $u(\tau) \in B^X(\overline{u}; r)$ for all $\tau \in [s, t]$. Then,*

$$\|u(t) - u(s)\|_{Z^*} \leq (D\delta\theta)^{-1}[\Phi(u(s)) - \Phi(\overline{u})]^\theta. \quad (2.32)$$

Proof For $s \leq \tau \leq t$, we have

$$\frac{d}{d\tau}[\Phi(u(\tau)) - \Phi(\overline{u})]^\theta = \theta[\Phi(u(\tau)) - \Phi(\overline{u})]^{\theta-1}\frac{d}{d\tau}\Phi(u(\tau)).$$

Then, (2.26) and (2.30) provide the inequality

$$-\frac{d}{d\tau}[\Phi(u(\tau) - \Phi(\overline{u})]^\theta \geq \delta\theta[\Phi(u(\tau)) - \Phi(\overline{u})]^{\theta-1}\|\dot{\Phi}(u(\tau))\|_Z \|u'(\tau)\|_{Z^*}.$$

Furthermore, the gradient inequality (2.31) provides the estimate

$$-\frac{d}{d\tau}[\Phi(u(\tau) - \Phi(\overline{u})]^\theta \geq D\delta\theta \|u'(\tau)\|_{Z^*}.$$

Integrating this inequality on $[s, t]$, we obtain that

$$[\Phi(u(s)) - \Phi(\overline{u})]^\theta - [\Phi(u(t)) - \Phi(\overline{u})]^\theta \geq D\delta\theta \int_s^t \|u'(\tau)\|_{Z^*} d\tau$$

$$\geq D\delta\theta \|u(t) - u(s)\|_{Z^*},$$

which in particular shows the desired estimate (2.32). $\qquad \square$

In view of (2.18) and (2.20), (2.32) immediately implies that

$$\|u(t) - u(s)\|_X \le C(2R)^{\frac{1}{2}} (D\delta\theta)^{-\frac{1}{2}} [\Phi(u(s)) - \Phi(\overline{u})]^{\frac{\theta}{2}}.$$

This estimate then yields a very strong result that once the trajectory $u(t)$ approaches sufficiently close to $\overline{u}$, the whole trajectory after that time stays in the ball $B^X(\overline{u}; r)$. In fact, fix an integer N so that $\|u(t_N) - \overline{u}\|_X \le \frac{r}{3}$ and that $C(2R)^{\frac{1}{2}} (D\delta\theta)^{-\frac{1}{2}} [\Phi(u(t_N)) - \Phi(\overline{u})]^{\frac{\theta}{2}} \le \frac{r}{3}$ due to (2.28). Then, by Proposition 2.4 (with $s = t_N$), if $u(t)$ is such that $u(\tau) \in B^X(\overline{u}; r)$ for all $\tau \in [t_N, t]$, then

$$\|u(t) - \overline{u}\|_X \le \|u(t) - u(t_N)\|_X + \|u(t_N) - \overline{u}\|_X$$

$$\le C(2R)^{\frac{1}{2}} (D\delta\theta)^{-\frac{1}{2}} [\Phi(u(t_N)) - \Phi(\overline{u})]^{\frac{\theta}{2}} + \|u(t_N) - \overline{u}\|_X \le \frac{2r}{3}.$$

This means that, after the time t_N, the trajectory cannot go out from the ball $\overline{B}^X(\overline{u}; \frac{2r}{3}) \subset B^X(\overline{u}; r)$. Moreover, (2.28) jointly with (2.29) yields that

$$\lim_{t \to \infty} \Phi(u(t)) = \inf_{t_N \le t < \infty} \Phi(u(t)) = \Phi(\overline{u}).$$

The main result now follows directly from those above.

Theorem 2.1 *Under the situations* (2.17), (2.19)–(2.22) *and* (2.24), *let an ω-limit $\overline{u}$ satisfy the structural conditions* (I), (II), (III) *and* (IV). *Then, as $t \to \infty$, $u(t)$ converges to $\overline{u}$ at the rate*

$$\|u(t) - \overline{u}\|_{Z^*} \le (D\delta\theta)^{-1} [\Phi(u(t)) - \Phi(\overline{u})]^{\theta} \qquad \text{for all} \quad t \ge t_N, \tag{2.33}$$

where t_N is the time mentioned above.

Proof Let s and t_n be such that $t_N \le s < t_n$. Apply Proposition 2.4 to $u(t_n) - u(s)$ to observe that

$$\|u(t_n) - u(s)\|_{Z^*} \le (D\delta\theta)^{-1} [\Phi(u(s)) - \Phi(\overline{u})]^{\theta}.$$

Then, letting $n \to \infty$, we obtain the desired estimate (2.33). $\qquad\qquad\square$

2.4 Comments for Applications

As stated, we have set up the frameworks and have made the structural assumptions all by taking account of the study on the typical equation (2.1). It may, however, be meaningful to explain how to choose the triplet $Z \subset X \subset Z^*$ and how to verify the four assumptions (I)–(IV) in a general context.

We suppose that the function $u(t)$ is usually given as a global solution of some evolution equation $u' = F(u)$, $0 \leq t < \infty$, where $F(u)$ is a continuous nonlinear operator from a Hilbert space Z into its adjoint space Z^*, Z being compactly embedded in Z^*. Therefore, $u(t)$ is naturally set as a function belonging to $\mathscr{C}([0, \infty); Z) \cap \mathscr{C}^1([0, \infty); Z^*)$ together with the boundedness (2.20). The existence of a Lyapunov function $\Phi(u)$ defined in a Hilbert space X, where the spaces $Z \subset X \subset Z^*$ make a triplet, is a major premise of our arguments. Along the trajectory of $u(t)$, it holds that $\frac{d}{dt}\Phi(u(t)) \leq 0$.

(I) *Critical Condition.* As (2.13), integration of $\frac{d}{dt}\Phi(u(t)) \leq 0$ for $t \in [0, \infty)$ yields integrability $\int_0^\infty \|u'(t)\|_{Z^*}^2 dt < \infty$. Therefore, there exists a temporal sequence $t_n \nearrow \infty$ such that $u'(t_n) \to 0$ in Z^*, i.e., $F(u(t_n)) \to 0$ in Z^*. Meanwhile, due to (2.20), we can assume that $u(t_n) \to \overline{u} \in \omega(u)$ in Z^* and $u(t_n) \to \overline{u}$ weakly in Z, which implies that $F(u(t_n)) \to F(\overline{u})$ weakly in Z^*. Hence, we have $F(\overline{u}) = 0$. But, as (2.10) shows, it is the case that $\dot{\Phi}(u) = 0$ if and only $F(u) = 0$. Thereby, $F(\overline{u}) = 0$ implies $\dot{\Phi}(\overline{u}) = \Phi'(\overline{u}) = 0$.

(II) *Lyapunov Function.* Assume that $\frac{d}{dt}\Phi(u(t)) = 0$ at some $t = \overline{t}$. Then, (2.26) jointed with the angle condition (2.30) yields $\dot{\Phi}(u(\overline{t})) = 0$ or $u'(\overline{t}) = 0$. As $\dot{\Phi}(u(\overline{t})) = 0$ yields $F(u(\overline{t})) = 0$, $u(\overline{t})$ is in any case a stationary solution and $u(t) \equiv u(\overline{t})$ for all $t \geq \overline{t}$, and the assertion of Theorem 2.1 is trivial. Therefore, it suffices to argue under the condition that $\frac{d}{dt}\Phi(u(t)) < 0$ for any $0 \leq t < \infty$, provided that the angle condition is fulfilled.

As mentioned, the existence of function $\Phi(u)$ satisfying $\frac{d}{dt}\Phi(u(t)) \leq 0$ for $0 \leq t < \infty$ is a major premise in our arguments.

(III) *Angle Condition.* This condition is more essential than the former two. Very roughly speaking, we can consider that the product $\langle \dot{\Phi}(u(t)), u'(t) \rangle_{Z \times Z^*}$ is given by $\|\dot{\Phi}(u(t))\|_Z \|u'(t)\|_{Z^*} \cos \Theta$, where Θ stands for the angle of two vectors $\dot{\Phi}(u(t))$ and $u'(t)$. Then, (2.30) yields $-\cos \Theta \geq \delta$, namely, Θ must be strictly larger that $\frac{\pi}{2}$. As seen by (2.10), the most favorable case is that the underlying evolution equations are of gradient form, namely, $u'(t) = F(u(t))$ is always in the direction of $-\dot{\Phi}(u(t))$, i.e., $\Theta = \pi$.

Anyway, (2.30) is verified by careful estimates on the derivative $-\frac{d}{dt}\Phi(u(t))$.

(IV) *Gradient Inequality.* As a matter of fact, this is the most crucial assumption. Consider the surface $\Phi = \Phi(u)$, $u \in Z$, in the product space $(u, \Phi) \in Z \times \mathbb{R}$. The critical condition $\Phi'(\overline{u}) = 0$ means that this surface is tangential to the horizontal plane $\Phi = \Phi(\overline{u})$ at $u = \overline{u}$. The gradient inequality then requires that its tangency must be in a moderate degree. In order to verify this inequality, we have to analyze precisely the behavior of $\dot{\Phi}(u) = J^{-1}\Phi'(u)$ in a neighborhood of $\overline{u}$.

The next chapter will, in fact, be devoted to investigating how to verify this inequality by the methods of Functional Analysis.

2.5 Notes and Future Studies

Consider a system of ordinary differential equations

$$\begin{cases} \xi' = -F_\xi(\xi, \eta), & 0 < t < \infty, \\ \eta' = -F_\eta(\xi, \eta), & 0 < t < \infty, \end{cases} \tag{2.34}$$

of divergence form for unknown functions $\xi = \xi(t)$ and $\eta = \eta(t)$. Here, $F(\xi, \eta)$ is a given function of class $\mathscr{C}^\infty$ for $(\xi, \eta) \in \mathbb{R}^2$.

For any (local or global) solution $\xi(t)$, $\eta(t)$ to (2.34), it holds that

$$\frac{d}{dt} F(\xi(t), \eta(t)) = F_\xi(\xi(t), \eta(t))\xi'(t) + F_\eta(\xi(t), \eta(t))\eta'(t)$$

$$= -\left([\xi'(t)]^2 + [\eta'(t)]^2\right) \le 0.$$

Therefore, $F(\xi, \eta)$ is a smooth Lyapunov function. It is, however, known that (2.34) may have a bounded global solution whose ω-limit set is a continuum. For instance, Palis–de Melo present in [PdeM82, Section 1] the following function:

$$F(\xi, \eta) \equiv F(r \cos \theta, r \sin \theta) = - \begin{cases} e^{1/(r^2-1)}, & 0 \le r < 1, \\ 0, & r = 1, \\ e^{1/(1-r^2)} \sin\left(\frac{1}{r-1} - \theta\right), & 1 < r < \infty, \end{cases} \tag{2.35}$$

where $\xi = r \cos \theta$, $\eta = r \sin \theta$ for $0 \le r < \infty$, $0 \le \theta < 2\pi$. Certainly, $F(\xi, \eta)$ is a bounded $\mathscr{C}^\infty$ function of $(\xi, \eta) \in \mathbb{R}^2$. Then, it is proved that there exists an initial value $1 + \frac{1}{2\pi} < \xi(0) < 1 + \frac{1}{\pi}$, $\eta(0) = 0$ for which the global solution has an ω-limit set of equilibria coinciding with the unit circle $\{(\xi, \eta); \xi^2 + \eta^2 = 1\}$.

On the basis of this example, we can immediately construct a counter example of infinite dimension showing that the existence of a smooth Lyapunov function does not necessarily guarantee the asymptotic convergence as in Theorem 2.1. Here we sketch its construction.

Let $\mathscr{X}$ be a separable Hilbert space with inner product $(\cdot, \cdot)$ and norm $\| \cdot \|$. Let A be a positive-definite self-adjoint operator of $\mathscr{X}$ with domain $\mathscr{D}(A)$ which is compactly embedded in $\mathscr{X}$. Let $0 < \lambda_1 \le \lambda_2 \le \cdots \le \lambda_n \le \cdots$ be its eigenvalues and let $u_1, u_2, \ldots, u_n, \ldots$ be their corresponding eigenvectors which compose an orthonormal basis of $\mathscr{X}$. Assume that among λ_n's, there is at least one multiple eigenvalue, say $\lambda > 0$, and let v_1, v_2 denote its two orthonormal eigenvectors. Using the function $F(\xi, \eta)$ given by (2.35), we will set a nonlinear operator

$$f(u) = \lambda u - [F_\xi((u, v_1), (u, v_2))v_1 + F_\eta((u, v_1), (u, v_2))v_2], \qquad u \in \mathscr{X},$$

and consider an evolution equation

$$u' + Au = f(u), \qquad 0 < t < \infty, \tag{2.36}$$

in $\mathscr{X}$. Let $u_0 \in \mathscr{D}(A)$ be such that $u_0 = \sum_{\lambda_n > \lambda} a_n u_n$ with coefficients a_n satisfying $\sum_{\lambda_n > \lambda} \lambda_n^2 |a_n|^2 < \infty$. If $\xi(t)$, $\eta(t)$ is a solution of (2.34), then

$$u(t) = e^{\lambda t} e^{-tA} u_0 + [\xi(t)v_1 + \eta(t)v_2]$$

becomes a solution to (2.36). Thereby, if $\xi(t)$, $\eta(t)$ is the global solution of (2.34) having an ω-limit set of the unit circle, then, since $e^{\lambda t} e^{-tA} u_0 = \sum_{\lambda_n > \lambda} e^{(\lambda - \lambda_n)t} a_n u_n \to 0$ in $\mathscr{X}$, $u(t)$ must similarly have an ω-limit set $\omega(u) = \{\overline{\xi} v_1 + \overline{\eta} v_2; \ \overline{\xi}^2 + \overline{\eta}^2 = 1\}$. Furthermore, take an inner product of the equation of (2.36) and $u'(t)$. Then,

$$\|u'(t)\|^2 + \frac{1}{2} \frac{d}{dt} \|A^{\frac{1}{2}} u(t)\|^2 = \frac{\lambda}{2} \frac{d}{dt} \|u(t)\|^2 - \frac{d}{dt} F((u(t), v_1), (u(t), v_2)).$$

This in turn means that the function

$$\Phi(u) = \frac{1}{2} \|A^{\frac{1}{2}} u\|^2 - \frac{\lambda}{2} \|u\|^2 + F((u, v_1), (u, v_2)), \qquad u \in \mathscr{D}(A^{\frac{1}{2}}),$$

is a Lyapunov function for the solution $u(t)$.

Taking account of these situations, set a triplet $Z \subset X \subset Z^*$ so that $Z = \mathscr{D}(A)$, $X = \mathscr{D}(A^{\frac{1}{2}})$ and $Z^* = \mathscr{X}$. Then, $u(t)$ belongs to (2.19) and satisfies (2.20). In the sense of (2.21)–(2.22), $\Phi : X \to \mathbb{R}$ is continuously Fréchet differentiable with the derivative

$$\dot{\Phi}(u) = u + A^{-1}[-\lambda u + F_\xi((u, v_1), (u, v_2))v_1 + F_\eta((u, v_1), (u, v_2))v_2], \qquad u \in X,$$

which also satisfies (2.24). In addition, it holds that

$$-\langle \dot{\Phi}(u(t)), u'(t) \rangle_{Z \times Z^*} = -\frac{d}{dt} \Phi(u(t)) = \|u'(t)\|_{Z^*}^2 = \|\dot{\Phi}(u(t))\|_Z^2.$$

Thus, the structural conditions (I), (II) and (III) are all fulfilled by any $\overline{u} \in \omega(u)$.

As a matter of fact, Theorem 2.1 is applicable for globally bounded solutions of various nonlinear parabolic evolution equations. In Chapters 3, 4 and 5 of vol. 2 [Yag], we shall handle some parabolic equations by following the comments mentioned in the preceding section. On the other hand, we have to remark that there are still inapplicable parabolic evolution equations which admit a global Lyapunov function. According to [Yag10, Chapter 11], Yagi–Primicerio [YP14] or Fusi–Primicerio–Yagi [FPY16], some equations in forest growth models have a Lyapunov function for global solutions but their ω-limit sets are seen to be nonempty only with respect to some weak topologies and it is still unknown whether those ω-limit sets are a singleton or not.

The methods explained in this chapter have already been used implicitly in the papers Azizi–Mola–Yagi [AMY17], Iwasaki–Yagi [IY] and Iwasaki–Osaki–Yagi [IOY] in which the authors established the asymptotic convergence for Laplace and reaction–diffusion equations, epitaxial growth equations and one-dimensional Keller–Segel equations, respectively. Actually, the general settings, the structural assumptions and the statements of Theorem 2.1 were all derived from the arguments on these specific parabolic equations.

Chapter 3
Łojasiewicz–Simon Gradient Inequality

In this chapter, let us investigate how to obtain the gradient inequality (2.31) in our general settings. But, before doing so, let us consider a favorable case in which (2.31) can be verified straightforwardly.

3.1 Favorable Case

Let X be a real Hilbert space with inner product $(\cdot, \cdot)_X$. Consider a functional $\Phi : X \to \mathbb{R}$ defined in the whole space X which is continuously differentiable in the sense of (2.21)–(2.22). That is, for each $f \in X$, it holds that,

$$\text{as } h \to 0 \text{ in } X, \quad \Phi(f + h) - \Phi(f) - (\dot{\Phi}(f), h)_X = o(\|h\|_X) \tag{3.1}$$

and that the mapping $u \mapsto \dot{\Phi}(u)$ is continuous from X into itself.

Let $\overline{u} \in X$ be a critical point of $\Phi(u)$, i.e., $\Phi'(\overline{u}) = J\dot{\Phi}(\overline{u}) = 0$. ($J : X \to X'$ is the isometric isomorphism defined by (1.2).)

Consider the case that the following two conditions are satisfied:

$$\text{the derivative } \dot{\Phi} : X \to X \text{ is Fréchet differentiable at } \overline{u}; \tag{3.2}$$

$$\text{its Fréchet derivative } L = [\dot{\Phi}]'(\overline{u}) \text{ is an isomorphism of } X. \tag{3.3}$$

Then, since

$$\Phi(u) - \Phi(\overline{u}) = \int_0^1 (\dot{\Phi}(\theta u + (1 - \theta)\overline{u}), u - \overline{u})_X d\theta$$

© The Author(s), under exclusive license to Springer Nature Singapore Pte Ltd. 2021
A. Yagi, *Abstract Parabolic Evolution Equations and Łojasiewicz-Simon Inequality I*, SpringerBriefs in Mathematics, https://doi.org/10.1007/978-981-16-1896-3_3

and since

$$\dot{\Phi}(\overline{u} + \theta(u - \overline{u})) = \theta[\dot{\Phi}]'(\overline{u})(u - \overline{u}) + r(\theta(u - \overline{u}))$$

due to (3.2), $r(h)$ being such that $\|r(h)\|_X = o(\|h\|_X)$ for small $h \in X$, it follows that, as $u \to \overline{u}$ in X,

$$|\Phi(u) - \Phi(\overline{u})| \leq [\|L\|_{\mathscr{L}(X)} + o(1)]\|u - \overline{u}\|_X^2.$$

Meanwhile, since

$$\dot{\Phi}(u) = [\dot{\Phi}]'(\overline{u})(u - \overline{u}) + r(u - \overline{u}),$$

it follows by (3.3) that, as $u \to \overline{u}$ in X,

$$\|\dot{\Phi}(u)\|_X \geq \|L(u - \overline{u})\|_X - o(\|u - \overline{u}\|_X) \geq [\|L^{-1}\|_{\mathscr{L}(X)}^{-1} - o(1)]\|u - \overline{u}\|_X.$$

Combine these estimates. As $\|L^{-1}\|_{\mathscr{L}(X)}^{-1} > 0$, if $\|u - \overline{u}\|_X$ is sufficiently small so that $o(1) < \|L^{-1}\|_{\mathscr{L}(X)}^{-1}$, then we obtain that

$$\|\dot{\Phi}(u)\|_X \geq c|\Phi(u) - \Phi(\overline{u})|^{\frac{1}{2}} \qquad \text{for all } u \in X \text{ sufficiently close to } \overline{u}. \qquad (3.4)$$

Because of (2.17), this yields the gradient inequality (2.31) with the optimal exponent $\theta = \frac{1}{2}$.

However, the assumptions (3.2) and (3.3) are sometimes too strong in the viewpoint of applications. For example, the Lyapunov function obtained for the Keller–Segel equations (see (5.29) of vol. 2) is not twice Fréchet differentiable, that is, (3.2) cannot be fulfilled.

Even when (3.2) is satisfied, (3.3) is unrealistic. For example, remember the Eq. (2.1). In this case, X was given by $X = \mathscr{D}(A^{\frac{1}{2}}) = \overset{\circ}{H}{}^1(\Omega)$ and $\dot{\Phi}(u)$ was given by $\dot{\Phi}(u) = u - A^{-1}f(u)$. The operator $u \mapsto \dot{\Phi}(u)$ is easily seen to be Fréchet differentiable at $\overline{u}$ with the derivative $Lh = h - A^{-1}f'(\overline{u})h$, $h \in X$ but cannot be expected to be an isomorphism of X.

Instead of (3.3), we will assume that L is a Fredholm operator of X. By definition, this assumption yields an orthogonal decomposition of X into $X = \mathscr{K}(L) + \mathscr{R}(L)$, where $\mathscr{K}(L)$ is a finite-dimensional subspace of X and $\mathscr{R}(L)$ is a closed subspace of X with finite codimension. Then, since L becomes an isomorphism from $\mathscr{R}(L)$ onto itself, arguments similar to those above can derive the gradient inequality of $\Phi(u)$ in $\mathscr{R}(L)$. On the other hand, since $\mathscr{K}(L)$ is finite-dimensional, we can apply the finite-dimensional gradient inequality due to Theorem 1.20 in $\mathscr{K}(L)$, assuming further that $\Phi(u)$ is analytic for $u \in \mathscr{K}(L)$.

3.2   General Settings and Structural Assumptions

Let X be a real Hilbert space with inner product $(\cdot, \cdot)_X$. Consider a functional $\Phi : X \to \mathbb{R}$ which is continuously differentiable in the sense of (2.21)–(2.22) with derivative $\dot{\Phi} : X \to X$. Let $\overline{u} \in X$ be a critical point of $\Phi(u)$, i.e.,

$$\Phi'(\overline{u}) = J\dot{\Phi}(\overline{u}) = 0. \tag{3.5}$$

We first assume that

$$\dot{\Phi} : X \to X \text{ is Gâteaux differentiable at } \overline{u} \tag{3.6}$$

and that its derivative $L = [\dot{\Phi}]'(\overline{u})$ (see Sect. 1.5) is not only a bounded linear operator from X into itself but also, more strongly,

$$L = [\dot{\Phi}]'(\overline{u}) \text{ is a Fredholm operator of } X. \tag{3.7}$$

It is possible to read (3.6)–(3.7) as conditions for the Fréchet derivative $\Phi' : X \to X'$. Because of $\dot{\Phi} = J^{-1}\Phi'$, where $J : X \to X'$ is the isometric isomorphism, (3.6) is equivalent to Gâteaux differentiability of $\Phi'(u)$ at $\overline{u}$. In addition, (3.7) is equivalent to say that its derivative $[\Phi']'(\overline{u})$ is not only a bounded linear operator from X into X' but also is a Fredholm operator.

Let us denote the kernel (resp. range) of L by $\mathcal{K}(L)$ (resp. $\mathcal{R}(L)$). By definition, $\mathcal{K}(L)$ is a finite-dimensional space of X and $\mathcal{R}(L)$ is a closed subspace of X having a finite codimension.

We next assume that there exists a real Banach space Y which is a linear subspace of X and satisfies the following five conditions:

$$Y \subset X \text{ with dense and continuous embedding;} \tag{3.8}$$

$$\overline{u} \in Y; \tag{3.9}$$

$$L^{-1}(Y) \subset Y, \text{ i.e., if } Lu \in Y, \text{ then } u \in Y; \tag{3.10}$$

$$\dot{\Phi} \text{ maps } Y \text{ into itself;} \tag{3.11}$$

$$\dot{\Phi} : Y \to Y \text{ is continuously Fréchet differentiable in } Y. \tag{3.12}$$

3.2.1   Some Properties of L in X

As for immediate properties of L in X, we can verify the following ones.

Proposition 3.1 *The operator L is symmetric in X, that is,*

$$(Lu, v)_X = (u, Lv)_X \quad \text{for any } u, v \in X. \tag{3.13}$$

Proof First, consider the case where u, $v \in Y$. Consider a function

$$\varphi(\theta_1, \theta_2) = \Phi(\overline{u} + \theta_1 u + \theta_2 v) - \Phi(\overline{u} + \theta_1 u) - \Phi(\overline{u} + \theta_2 v) + \Phi(\overline{u})$$

for (θ_1, θ_2) which varies in a small square $[0, \varepsilon]^2$. Then, by the mean value theorem applied for the variable θ_2, there is a number $\varepsilon' \in (0, \varepsilon)$ for which it holds that

$$\varphi(\varepsilon, \varepsilon) - \varphi(\varepsilon, 0) = \varphi_{\theta_2}(\varepsilon, \varepsilon')\varepsilon = (\dot{\Phi}(\overline{u} + \varepsilon u + \varepsilon' v), \varepsilon v)_X - (\dot{\Phi}(\overline{u} + \varepsilon' v), \varepsilon v)_X.$$

Since $\varphi(\varepsilon, 0) = 0$, it follows that

$$\varphi(\varepsilon, \varepsilon) = (\dot{\Phi}(\overline{u} + \varepsilon u + \varepsilon' v) - \dot{\Phi}(\overline{u} + \varepsilon' v), \varepsilon v)_X.$$

Then, the assumption (3.11) yields that

$$\varphi(\varepsilon, \varepsilon) = (\dot{\Phi}(\overline{u} + \varepsilon u + \varepsilon' v) - \dot{\Phi}(\overline{u} + \varepsilon' v) - [\dot{\Phi}]'(\overline{u} + \varepsilon' v)(\varepsilon u), \varepsilon v)_X$$

$$+ (\{[\dot{\Phi}]'(\overline{u} + \varepsilon' v) - [\dot{\Phi}]'(\overline{u})\}(\varepsilon u), \varepsilon v)_X + ([\dot{\Phi}]'(\overline{u})(\varepsilon u), \varepsilon v)_X$$

$$= (r_1(\varepsilon u), \varepsilon v)_X + \varepsilon(r_2(\varepsilon' v), \varepsilon v)_X + \varepsilon^2(Lu, v)_X,$$

where $r_1(\varepsilon u)$ and $r_2(\varepsilon' v)$ stand for suitable vectors of Y satisfying $\|r_1(\varepsilon u)\|_Y = o(\varepsilon)$ and $\|r_2(\varepsilon' v)\|_Y = o(1)$ as $\varepsilon \geq \varepsilon' \to 0$.

By similar arguments, we can obtain that

$$\varphi(\varepsilon, \varepsilon) = (r_1(\varepsilon v), \varepsilon u)_X + (r_2(\varepsilon' u), \varepsilon u)_X + \varepsilon^2(Lv, u)_X.$$

It therefore follows that $|(Lu, v)_X - (u, Lv)_X| \leq \varepsilon^{-1}o(\varepsilon)$. Letting $\varepsilon \to 0$, we conclude that (3.13) holds true for any pair of u and v from Y.

Next, consider the general case where $u, v \in X$. Due to (3.8), there are sequences u_n and v_n which converge to u and v in X respectively. As $(Lu_n, v_n)_X = (u_n, Lv_n)_X$ and as Lu_n and Lv_n converge to Lu and Lv respectively, we verify that (3.13) holds true for any pair of u, $v \in X$. $\square$

As $\mathcal{K}(L)$ is finite-dimensional, $\mathcal{K}(L)$ is a closed subspace of X. Denote by P the orthogonal projection from X onto $\mathcal{K}(L)$. By definition, we have $LP = 0$ on X. But, since P is also symmetric in X, it follows that $PL = (LP)^* = 0$, i.e.,

$$LP = PL = 0 \quad \text{on} \quad X. \tag{3.14}$$

The projection P naturally induces an orthogonal decomposition of X into $X = PX + (I - P)X$, where $PX = (I - P)^{-1}0 = \mathcal{K}(L)$ and $(I - P)X = P^{-1}0$.

Proposition 3.2 *The space X is orthogonally decomposed into*

$$X = \mathcal{K}(L) + L(X), \tag{3.15}$$

its projections being given by P and $I - P$ respectively. Consequently, L is an isomorphism from $L(X)$ onto itself.

Proof Let us verify that $(I - P)X = L(X)$. If $v \in L(X)$, then $v = Lu$ with some $u \in X$; therefore, $Pv = PLu = 0$; so that, $v = (I - P)v \in (I - P)X$. Hence, $L(X) \subset (I - P)X$. Suppose that $L(X) \subsetneq (I - P)X$. Then, as $L(X)$ is a closed subspace of $(I - P)X$, there must exist a vector $0 \neq u^* \in (I - P)X$ such that $(Lu, u^*) = (u, Lu^*) = 0$ for any $u \in X$; therefore, $Lu^* = 0$ and $u^* \in \mathcal{K}(L)$, i.e., $Pu^* = u^*$. Meanwhile, let $u^* = (I - P)v^*$; then, $0 = (I - P)u^* = (I - P)^2 v^* = (I - P)v^* = u^*$, which is a contradiction. Hence, $L(X) = (I - P)X$.

It is clear that L is a bijection from $L(X)$ onto itself. Then, thanks to a corollary of the open mapping theorem (see [Yos80, p. 77]), L is concluded to be an isomorphism of $L(X)$. □

3.2.2 Some Properties of L in Y

Since $\dot{\Phi}$ is Gâteaux differentiable at $\overline{u}$ in X and since $\dot{\Phi}$ is Fréchet differentiable in Y, we naturally have $[\dot{\Phi}]'(\overline{u})h = Lh$ for $h \in Y$. Because $[\dot{\Phi}]'(\overline{u}) \in \mathcal{L}(Y)$, this means that L is also a bounded linear operator of Y.

Notice that (3.10) implies that $\mathcal{K}(L) = L^{-1}0 \subset L^{-1}Y \subset Y$. Therefore, $\mathcal{K}(L)$ is also a finite-dimensional subspace of Y. In particular, $\mathcal{K}(L)$ is also a closed subspace of Y. In addition, since the two norms $\| \cdot \|_Y$ and $\| \cdot \|_X$ are equivalent on the space $\mathcal{K}(L)$, we see that $\|Pu\|_Y \leq C\|Pu\|_X \leq C\|u\|_X$, which means that P is a bounded operator from Y into itself. Therefore, we have $P \in \mathcal{L}(Y)$ and $P^2 = P$.

Consequently, P also induces a topological decomposition of Y, in the manner that $Y = PY + (I - P)Y$. Since $PY = (I - P)^{-1}0$ and $(I - P)Y = P^{-1}0$, PY and $(I - P)Y$ are a closed subspace of Y.

Proposition 3.3 *It is seen that $PY = \mathcal{K}(L)$ and $(I - P)Y = L(Y)$. Consequently, Y is represented as a topological direct sum of the form*

$$Y = \mathcal{K}(L) + L(Y). \tag{3.16}$$

Moreover, L is an isomorphism from $L(Y)$ onto itself.

Proof As P was a projection of Y onto $\mathcal{K}(L)$, it is clear that $PY = \mathcal{K}(L)$.

Let us verify that $(I - P)Y = L(Y)$. If $v \in (I - P)Y$, then v must be in $L(X)$ due to (3.15); namely, $v = Lu$ with some $u \in X$; but (3.10) implies that $u \in Y$; hence, $v \in L(Y)$. Conversely, if $v \in L(Y)$, then $v = Lu$ with some $u \in Y$; therefore, $Pv = PLu = 0$ from (3.14); hence, $v = (I - P)v \in (I - P)Y$.

In this way, Y has been verified to be topologically decomposed as in (3.16).

Finally, let us verify that $L : L(Y) \rightarrow L(Y)$ is an isomorphism. But this is concluded by the same arguments as in the proof of Proposition 3.2, for it is clear that L is a bounded linear operator from $L(Y)$ into itself and is a bijection. □

3.3   Critical Manifold

By (3.12), we have $\dot{\Phi}(u) = [\dot{\Phi}]'(\overline{u})(u - \overline{u}) + r(u - \overline{u})$ for $u \in Y, r(u - \overline{u})$ being a vector of Y such that $\|r(u - \overline{u})\|_Y = o(\|u - \overline{u}\|_Y)$ as $u \to \overline{u}$ in Y. Therefore, if $\dot{\Phi}(u)$ is in $\mathscr{K}(L)$, i.e., $P\dot{\Phi}(u) = \dot{\Phi}(u)$, then we have $\dot{\Phi}(u) = P(r(u - \overline{u}))$ because of (3.14). This suggests that we cannot use analogous arguments as for (3.4) for estimating the norm $\|\dot{\Phi}(u)\|_Y$ by utilizing the property that L is an isomorphism of $L(Y)$.

We are therefore led to distinguish vectors $u \in Y$ for which $\dot{\Phi}(u)$ lie in $\mathscr{K}(L)$. (See Chill [Chi06].)

3.3.1   Definition

Let us call a surface

$$S = \{u \in Y;\ \dot{\Phi}(u) \in \mathscr{K}(L)\} = \{u \in Y;\ (I - P)\dot{\Phi}(u) = 0\} \tag{3.17}$$

of Y a critical manifold of $\Phi(u)$. As $\dot{\Phi}(\overline{u}) = 0$ and $\overline{u} \in Y$, $\overline{u}$ is certainly on S.

In view of (3.16), it is possible to identify Y as the product space $Y = \mathscr{K}(L) \times L(Y)$. Then, in a suitable neighborhood of $\overline{u}$, S can be represented as follows.

Proposition 3.4 *There is an open neighborhood $U = U_0 \times U_1$ of $\overline{u}$ in Y, where U_0 (resp. U_1) is an open neighborhood of $P\overline{u}$ (resp. $(I - P)\overline{u}$) in $\mathscr{K}(L)$ (resp. $L(Y)$), such that S is given in U by*

$$S \cap U = \{(u_0, g(u_0));\ u_0 \in U_0,\ g : U_0 \to U_1\},$$

where g is a mapping from U_0 into U_1 satisfying $g(P\overline{u}) = (I - P)\overline{u}$ and is continuously Fréchet differentiable.

Proof We employ the implicit function theorem. Let $G : U_0 \times U_1 \to L(Y)$ be a mapping defined by $G(u_0, u_1) = (I - P)\dot{\Phi}(u_0 + u_1)$ for $(u_0, u_1) \in U_0 \times U_1$. By (3.12), G is continuously differentiable in U and $G(P\overline{u}, (I - P)\overline{u}) = 0$. Proposition 3.3 provides that $G_{u_1}(P\overline{u}, (I - P)\overline{u}) = (I - P)L = L$ is an isomorphism from $L(Y)$ onto itself. Therefore, thanks to the implicit function theorem, if U_0 is replaced by a smaller one (if necessary), there exists a $\mathscr{C}^1$ mapping $g(u_0)$ for $u_0 \in U_0$ with values in U_1 such that $G(u_0, g(u_0)) = 0$ for any $u_0 \in U_0$ together with $g(P\overline{u}) = (I - P)\overline{u}$.   □

3.3.2 Decomposition of Y into $S + L(Y)$

Using the mapping $g : U_0 \rightarrow U_1$, we introduce another decomposition for the vectors of U (instead of (3.16)). In fact, for $u \in U$, u can be expressed by the form

$$u = [Pu + g(Pu)] + [(I - P)u - g(Pu)] \equiv u_S + u_1 \in S + L(Y), \qquad u \in U, \tag{3.18}$$

here $u_S = Pu + g(Pu) = (Pu, g(Pu)) \in S$ and $u_1 = (I - P)u - g(Pu) \in L(Y)$ respectively. We also notice that

$$\text{as } u \rightarrow \overline{u} \text{ in } Y, u_S \text{ (resp. } u_1) \text{ converges to } \overline{u} \text{ (resp. 0) in } Y. \tag{3.19}$$

This decomposition then provides the following important estimate for $\dot{\Phi}(u)$.

Proposition 3.5 *It holds that*

$$\|P\dot{\Phi}(u)\|_Y \geq \|\dot{\Phi}(u_S)\|_Y - o(1)\|u_1\|_Y, \qquad u = u_S + u_1 \in U, \tag{3.20}$$

$$\|(I - P)\dot{\Phi}(u)\|_Y \geq c\|u_1\|_Y, \qquad u = u_S + u_1 \in U, \tag{3.21}$$

provided that U is a sufficiently small neighborhood of $\overline{u}$ in Y. Here, $o(1)$ denotes a small quantity tending to 0 as $u \rightarrow \overline{u}$ in Y, while c is a fixed positive constant.

Proof We write

$$\dot{\Phi}(u) = \dot{\Phi}(u_S) + [\dot{\Phi}]'(u_S)(u - u_S) + r(u - u_S)$$

$$= \dot{\Phi}(u_S) + \{[\dot{\Phi}]'(u_S) - [\dot{\Phi}]'(\overline{u})\}u_1 + [\dot{\Phi}]'(\overline{u})u_1 + r(u_1),$$

where $\|r(u_1)\|_Y = o(\|u_1\|_Y)$. As $(I - P)\dot{\Phi}(u_S) = 0$ and $P[\dot{\Phi}]'(\overline{u})u_1 = 0$, we have

$$P\dot{\Phi}(u) = \dot{\Phi}(u_S) + P[[\dot{\Phi}]'(u_S) - [\dot{\Phi}]'(\overline{u})]u_1 + Pr(u_1).$$

Therefore,

$$\|P\dot{\Phi}(u)\|_Y \geq \|\dot{\Phi}(u_S)\|_Y - o(1)\|u_1\|_Y - o(\|u_1\|_Y),$$

where $o(1) = \|[\dot{\Phi}]'(u_S) - [\dot{\Phi}]'(\overline{u})\|_{\mathscr{L}(Y)}$ due to (3.12). Therefore, (3.20) is observed.

On the other hand, for the same reasons, we have

$$(I - P)\dot{\Phi}(u) = (I - P)[[\dot{\Phi}]'(u_S) - [\dot{\Phi}]'(\overline{u})]u_1 + [\dot{\Phi}]'(\overline{u})u_1 + (I - P)r(u_1).$$

Since Proposition 3.3 yields the estimate

$$\|u_1\|_{L(Y)} = \|L^{-1}Lu_1\|_{L(Y)} \le \|L^{-1}\|_{\mathscr{L}(L(Y))}\|[\dot{\Phi}]'(\overline{u})u_1\|_{L(Y)},$$

it follows that

$$\|(I - P)\dot{\Phi}(u)\|_Y \ge \|L^{-1}\|^{-1}\|u_1\|_Y - o(1)\|u_1\|_Y - o(\|u_1\|_Y).$$

Let us take a sufficiently small neighborhood U. Then, as u_S's (resp. u_1's) become close to $\overline{u}$ (resp. small) in Y due to (3.19), we can obtain (3.21).   □

Combining (3.20) and (3.21), we verify that

$$\|\dot{\Phi}(u)\|_Y \ge C\big[\|P\dot{\Phi}(u)\|_Y + \|(I - P)\dot{\Phi}(u)\|_Y\big]$$

$$\ge C\big[\|\dot{\Phi}(u_S)\|_Y + (c - o(1))\|u_1\|_Y\big].$$

Hence, by virtue of (3.19), it follows that

$$\|\dot{\Phi}(u)\|_Y \ge c[\|\dot{\Phi}(u_S)\|_Y + \|u_1\|_Y], \qquad u = u_S + u_1 \in U, \tag{3.22}$$

provided that U is a sufficiently small neighborhood of $\overline{u}$ in Y, where c is a fixed positive constant.

3.4 Gradient Inequality in $L(Y)$

Utilizing the decomposition (3.18) and the estimate (3.22), let us estimate the difference $|\Phi(u) - \Phi(\overline{u})|$ for the vectors $u \in U$. Estimating the difference by

$$|\Phi(u) - \Phi(\overline{u})| \le |\Phi(u) - \Phi(u_S)| + |\Phi(u_S) - \Phi(\overline{u})|, \qquad u = u_S + u_1 \in U,$$

we intend to estimate the two differences in the right hand side by $\|\dot{\Phi}(u)\|_Y$ but by quite different ways. This section is devoted to estimating the former difference.

By the assumption (3.12), we have

$$\Phi(u) - \Phi(u_S) = (\dot{\Phi}(u_S), u - u_S)_X + \tfrac{1}{2}([\dot{\Phi}]'(u_S)(u - u_S), u - u_S)_X + o(\|u - u_S\|_Y^2)$$

$$= (\dot{\Phi}(u_S), u_1)_X + \tfrac{1}{2}([\dot{\Phi}]'(u_S)u_1, u_1)_X + o(\|u_1\|_Y^2).$$

Therefore,

$$|\Phi(u) - \Phi(u_S)| \le \|\dot{\Phi}(u_S)\|_X\|u_1\|_X + C\|u_1\|_Y\|u_1\|_X + o(\|u_1\|_Y^2)$$

$$\le \|\dot{\Phi}(u_S)\|_X^2 + C\|u_1\|_Y^2.$$

By virtue of (3.22), we conclude that

$$\|\dot{\Phi}(u)\|_Y \geq C|\Phi(u) - \Phi(u_S)|^{\frac{1}{2}}, \qquad \forall u \in U, \tag{3.23}$$

where U is the same neighborhood as in (3.22).

3.5 Gradient Inequality on S

Let us next estimate $|\Phi(u_S) - \Phi(\overline{u})|$. As seen by Proposition 3.4, u_S is described as $u_S = u_0 + g(u_0)$, where $u_0 \in U_0 \subset \mathcal{K}(L)$ and $g(u_0) \in U_1 \subset L(Y)$.

Let $v_1, v_2, \ldots, v_N$ be a basis of $\mathcal{K}(L)$, where $N = \dim \mathcal{K}(L)$, and identify $\mathcal{K}(L)$ with $\mathbb{R}^N$ by the correspondence

$$u_0 = \sum_{k=1}^{N} \xi_k v_k \in \mathcal{K}(L) \quad \longleftrightarrow \quad \boldsymbol{\xi} = (\xi_1, \xi_2, \ldots, \xi_N) \in \mathbb{R}^N.$$

Let $P\overline{u} \leftrightarrow \overline{\boldsymbol{\xi}}$ and let U_0 correspond to an open neighborhood $\boldsymbol{\Omega}$ of $\overline{\boldsymbol{\xi}}$ in $\mathbb{R}^N$. We make the crucial assumption here that

the function $\boldsymbol{\xi} \in \boldsymbol{\Omega} \mapsto \phi(\boldsymbol{\xi}) \equiv \Phi\left(\sum_{k=1}^{N} \xi_k v_k + g(\sum_{k=1}^{N} \xi_k v_k)\right)$

$$\text{is analytic in a neighborhood of } \overline{\boldsymbol{\xi}}. \tag{3.24}$$

Then, the classical Theorem 1.20 provides that, for some exponent $0 < \theta \leq \frac{1}{2}$, $\phi(\boldsymbol{\xi})$ satisfies the gradient inequality

$$\|\nabla_{\boldsymbol{\xi}} \phi(\boldsymbol{\xi})\|_{\mathbb{R}^N} \geq D_0 |\phi(\boldsymbol{\xi}) - \phi(\overline{\boldsymbol{\xi}})|^{1-\theta}, \qquad \boldsymbol{\xi} \in \boldsymbol{\Omega},$$

with some constant $D_0 > 0$, provided that $\boldsymbol{\Omega}$ is replaced by a smaller one.

Meanwhile, in view of (3.1), (3.15) and the Fréchet differentiability of $g(u_0)$, we observe that

$$D_{\xi_k} \phi(\boldsymbol{\xi}) = (\dot{\Phi}(u_0 + g(u_0)), v_k + g'(u_0)(v_k))_X.$$

Therefore, $|D_{\xi_k} \phi(\boldsymbol{\xi})| \leq C\|\dot{\Phi}(u_0 + g(u_0))\|_X$. Remembering that $u_S = u_0 + g(u_0)$, $\phi(\boldsymbol{\xi}) = \Phi(u_S)$ and $\phi(\overline{\boldsymbol{\xi}}) = \Phi(\overline{u})$, we conclude that

$$\|\dot{\Phi}(u_S)\|_X \geq C|\Phi(u_S) - \Phi(\overline{u})|^{1-\theta}, \qquad u_S = u_0 + g(u_0), \ u_0 \in U_0,$$

provided that U_0 is replaced by a smaller one.

According to (3.22), $\|\dot{\Phi}(u)\|_Y \geq C\|\dot{\Phi}(u_S)\|_Y$ for $u \in U$ if U is a sufficiently small neighborhood of $\overline{u}$ in Y. Hence, it follows that

$$\|\dot{\Phi}(u)\|_Y \geq C|\Phi(u_S) - \Phi(\overline{u})|^{1-\theta}, \qquad \forall u \in U. \tag{3.25}$$

3.6   Main Theorem

We can now state the main theorem of this chapter.

Theorem 3.1 *Let* $\Phi : X \to \mathbb{R}$ *be a continuously differentiable functional in the sense of* (2.21)–(2.22) *with derivative* $\dot{\Phi} : X \to X$, *and let* $\overline{u}$ *be its critical point (as* (3.5)*). Assume that* $u \mapsto \dot{\Phi}(u)$ *is Gâteaux differentiable at* $\overline{u}$ *(as* (3.6)*) with derivative* L *satisfying* (3.7). *Assume also that there exists a Banach space* Y *satisfying* (3.8)–(3.12). *Define the critical manifold* S *as* (3.17) *and assume that the analyticity condition* (3.24) *is satisfied. Then, in a neighborhood* U *of* $\overline{u}$ *in* Y, $\Phi(u)$ *satisfies the gradient inequality*

$$\|\dot{\Phi}(u)\|_Y \geq C|\Phi(u) - \Phi(\overline{u})|^{1-\theta}, \qquad \forall u \in U, \tag{3.26}$$

where $0 < \theta \leq \frac{1}{2}$ *is the exponent given in* (3.25).

Proof It follows from $0 < \theta \leq \frac{1}{2}$ that $1 < \frac{1}{1-\theta} \leq 2$. In view of this, we combine the two estimates (3.23) and (3.25) to obtain that

$$|\Phi(u) - \Phi(\overline{u})| \leq |\Phi(u) - \Phi(u_S)| + |\Phi(u_S) - \Phi(\overline{u})|$$

$$\leq C[\|\dot{\Phi}(u)\|_Y^2 + \|\dot{\Phi}(u)\|_Y^{\frac{1}{1-\theta}}] \leq C\|\dot{\Phi}(u)\|_Y^{\frac{1}{1-\theta}}.$$

Hence, (3.26) has been derived.   □

3.7   Gradient Inequality with Respect to $\|\cdot\|_Z$

Let us return to the general settings stated in Sect. 3.2.2. What we have to verify is the gradient inequality (2.31) with respect to the norm of Z. Therefore, it still remains to consider a question of how we derive (2.31) from (3.26) which has just been established above.

3.7.1   Case Where $Z \subset Y$

Consider first the case where Y includes Z. More precisely, Y is such that $Z \subsetneqq Y \subset X$ and that the intermediate inequality

$$\|u\|_Y \leq C\|u\|_Z^\alpha \|u\|_X^{1-\alpha}, \qquad u \in Z, \tag{3.27}$$

holds with some $0 \leq \alpha < 1$. Let U be the neighborhood of $\bar{u}$ in Y stated in Theorem 3.1. Then, there exists $r''' > 0$ for which we have $B^X(\bar{u}; r''') \cap \overline{B}^Z(0; R) \subset U$, where R is the radius appearing in (2.20). Since $\|\dot{\Phi}(u)\|_Z \geq C\|\dot{\Phi}(u)\|_Y$ for $u \in Z$, (3.26) can yield the desired inequality (2.31).

3.7.2   Case Where $Y \subset Z$

In this case, (3.26) cannot yield (2.31) in any direct way. So, we need to introduce another supplementary Banach space $W \subset Y$ and to assume a global uniform boundedness of $u(t)$ with respect to the W-norm.

More precisely, let W denote a Banach space such that $W \subset Y \subset Z$. Assume that, among the spaces $W \subset Y \subset X$, the intermediate inequality

$$\|u\|_Y \leq C\|u\|_W^\alpha \|u\|_X^{1-\alpha}, \qquad u \in W, \tag{3.28}$$

holds with some $0 < \alpha < 1$ and that, among the spaces $W \subset Y \subset Z$, the estimate

$$\|u\|_Y \leq C\|u\|_W^\beta \|u\|_Z^{1-\beta}, \qquad u \in W, \tag{3.29}$$

holds with some $0 \leq \beta < \theta$, θ being the exponent appearing in (3.26).

Meanwhile, assume that the solution $u(t)$ satisfies the global norm estimate

$$\|u(t)\|_W \leq R', \qquad 0 \leq t < \infty, \tag{3.30}$$

with some constant $R' > 0$ and that $u \mapsto \dot{\Phi}(u)$ is a mapping from W into itself together with the estimate

$$\|\dot{\Phi}(u)\|_W \leq C, \qquad u \in \overline{B}^W(0; R'). \tag{3.31}$$

It then follows by (3.28) that there exists $r''' > 0$ such that

$$B^X(\bar{u}; r''') \cap \overline{B}^W(0; R') \subset U,$$

where U is the neighborhood of $\overline{u}$ in Y stated in Theorem 3.1. On the other hand, by (3.29) and (3.31), we have

$$C\|\dot{\Phi}(u)\|_Z^{1-\beta} \geq \|\dot{\Phi}(u)\|_Y, \qquad u \in \overline{B}^W(0; R').$$

Therefore, (3.26) yields the estimate

$$\|\dot{\Phi}(u)\|_Z \geq D'|\Phi(u) - \Phi(\overline{u})|^{1-\theta'}, \qquad u \in B^X(\overline{u}; r''') \cap \overline{B}^W(0; R'), \qquad (3.32)$$

with an exponent $\theta' = \frac{\theta-\beta}{1-\beta}$ which is positive because of $\beta < \theta$.

Thus, (3.32) together with (3.30) can yield the desired inequality (2.31).

3.8   Notes and Future Studies

As reviewed in Chap. 1 (Theorem 1.20), the gradient inequality (2.31) was first proved by Łojasiewicz [Loj63, Loj65] in a domain $\Omega \subset \mathbb{R}^N$ under the assumption that $\phi(\xi)$ is analytic for $\xi \in \Omega$. On the other hand, we know that the $\mathscr{C}^\infty$ function $F(\xi, \eta)$ given by (2.35) satisfies (2.31) nowhere on the unit circle, all the points of which are a critical point of $F(\xi, \eta)$, because, otherwise, Theorem 2.1 must be available to the solutions of the Eq. (2.34). This means that some moderateness of degree of degeneration like analyticity is necessary in order that the gradient inequality takes place at the critical points.

After 20 years, the inequality was extended by Simon [Sim83] into the infinite-dimensional space, that is, $\Phi : O \to \mathbb{R}$ is a suitable function defined in an open subset O of a Banach space X. Afterward, Chill, Haraux, Jendoubi and many other researchers developed Simon's methods and applied their results to various partial differential equations in papers including [Che08, Che09, Chi03, CHJ09, HJ98, HJ99, HJ07, Jen98]. Overviews of these results and systematic treatments were made in the monograph by Huang [Hua06].

Furthermore, Chill [Chi06] and Haraux–Jendoubi [HJ11] presented the general methods in the framework of Functional Analysis of obtaining the inequality. In [HJ11], they showed that for a function $\Phi : X \to \mathbb{R}$ (X being a Hilbert space) the conditions that $\Phi : X \to \mathbb{R}$ is analytic and that $L = [\dot{\Phi}]'(\overline{u}) : X \to X$ is a Fredholm operator can provide the inequality at the critical point $\overline{u}$. In [Chi06], he pointed out that the analyticity of $\Phi : X \to \mathbb{R}$ on the whole space X can be relaxed into $\mathscr{C}^2$-regularity on X and analyticity only on the finite-dimensional critical manifold given by $S = \{u \in X; \ \dot{\Phi}(u) \in \mathscr{K}(L)\}$.

Meanwhile, Feireisl–Issard-Roch–Petzeltová [FIP04] devised a non-smooth version of Simon's methods by using the theory of maximal monotone operators in which $\Phi' : X \to X'$ is not required to be differentiable. Their results seem to be much more well-adapted to some types of partial differential equations, see [FIP04] and Feireisl–Laurençot–Petzeltová [FLP07].

In this chapter, we followed the same spirit as Chill [Chi06] and Haraux–Jendoubi [HJ11] and modified their results. The main difference is that we introduced, in addition to $Z \subset X \subset Z^*$, another Banach space Y which satisfies (3.8)–(3.12). Thanks to this space Y, $\dot{\Phi} : X \to X$ is allowed to be simply Gâteaux differentiable at $\overline{u}$ and the analyticity of $\Phi : S \to \mathbb{R}$ is allowed to be so with respect to the topology of Y. As will be seen in Subsections 3.6.2, 4.6.2 and 5.6.2 of vol. 2 [Yag], when the triplet $Z \subset X \subset Z^*$ is set according to Comments for Applications of Chap. 2, finding the space Y is done in an obvious way.

Bibliography

[AMY17] S. Azizi, G. Mola, A. Yagi, Longtime convergence for epitaxial growth model under Dirichlet conditions. Osaka J. Math. **54**, 689–706 (2017)

[Bre11] H. Brezis, *Functional Analysis, Sobolev Spaces and Partial Differential Equations* (Springer, Berlin, 2011)

[Che08] L. Chergui, Convergence of global and bounded solutions of a second order gradient like system with nonlinear dissipation and analytic nonlinearity. J. Dyn. Differ. Equ. **20**, 643–652 (2008)

[Che09] L. Chergui, Convergence of global and bounded solutions of the wave equation with nonlinear dissipation and analytic nonlinearity. J. Evol. Equ. **9**, 405–418 (2009)

[Chi03] R. Chill, On the Łojasiewicz-Simon gradient inequality. J. Funct. Anal. **201**, 572–601 (2003)

[Chi06] R. Chill, The Lojasiewicz-Simon gradient inequality on Hilbert spaces, in *Proceedings of the 5th European-Maghrebian Workshop on Semigroup Theory, Evolution Equations and Applications* (2006), pp. 25–36

[CHJ09] R. Chill, A. Haraux, M.A. Jendoubi, Applications of the Łojasiewicz-Simon gradient inequality to gradient-like evolution equations. Anal. Appl. **7**, 351–372 (2009)

[DL84a] R. Dautray, J.L. Lions, *Analyse mathématique et calcul numérique pour les sciences et les techniques,* vol. 1 (Masson, Paris, 1984)

[DL84b] R. Dautray, J.L. Lions, *Analyse mathématique et calcul numérique pour les sciences et les techniques,* vol. 2 (Masson, Paris, 1984)

[FIP04] E. Feireisl, F. Issard-Roch, H. Petzeltová, A non-smooth version of the Lojasiewicz-Simon theorem with applications to non-local phase-field systems. J. Differ. Equ. **199**, 1–21 (2004)

[FLP07] E. Feireisl, P. Laurençot, H. Petzeltová, On convergence to equilibria for the Keller-Segel chemotaxis model. J. Differ. Equ. **236**, 551–569 (2007)

[FPY16] L. Fusi, M. Primicerio, A. Yagi, A mathematical model for forest growth dynamics. J. Math. Anal. Appli. **440**, 773–793 (2016)

[HJ98] A. Haraux, M.A. Jendoubi, Convergence of solutions of second-order gradient-like systems with analytic nonlinearities. J. Differ. Equ. **144**, 313–320 (1998)

[HJ99] A. Haraux, M.A. Jendoubi, Convergence of bounded weak solutions of the wave equation with dissipation and analytic nonlinearity. Calc. Var. Partial Differ. Equ. **9**, 95–124 (1999)

[HJ07] A. Haraux, M.A. Jendoubi, On the convergence of global and bounded solutions of some evolution equations. J. Evol. Equ. **7**, 449–470 (2007)

© The Author(s), under exclusive license to Springer Nature Singapore Pte Ltd. 2021
A. Yagi, *Abstract Parabolic Evolution Equations and Łojasiewicz-Simon Inequality I,*
SpringerBriefs in Mathematics, https://doi.org/10.1007/978-981-16-1896-3

[HJ11]   A. Haraux, M.A. Jendoubi, The Łojasiewicz gradient inequality in the infinite-
            dimensional Hilbert space framework. J. Funct. Anal. **260**, 2826–2842 (2011)
[Hua06]   S.-Z. Huang, *Gradient Inequalities with Applications to Asymptotic Behavior and
            Stability of Gradient-like Systems*. Mathematical Surveys Monographs, vol. 126 (AMS,
            Providence, 2006)
[IOY]   S. Iwasaki, K. Osaki, A. Yagi, Asymptotic convergence of solutions for one-
            dimensional Keller-Segel equations, http://hdl.handle.net/11094/77680
[IY]   S. Iwasaki, A. Yagi, Asymptotic convergence of solutions for Laplace reaction-
            diffusion equations. Nonlinear Anal. RWA **51**, 102986 (2020)
[Jen98]   M.A. Jendoubi, A simple unified approach to some convergence theorem by L. Simon.
            J. Funct. Anal. **153**, 187–202 (1998)
[LM68]   J.L. Lions, E. Magenes, *Problème aux limites non homogènes et applications,* vol. 1
            (Dunord, Paris, 1968)
[Loj63]   S. Łojasiewicz, Une propriété topologique des sous-ensembles analytique réels. Col-
            loques internationaux du C.N.R.S.: *Les équations aux dérivées partielles, Paris, 1962*
            (Editions du C.N.R.S., Paris, 1963), pp. 87–89
[Loj65]   S. Łojasiewicz, *Ensembles Semi-Analytiques* (Publ. Inst. Hautes Etudes Sci., Bures-
            sur-Yvette, 1965)
[LZ99]   S. Łojasiewicz, M.A. Zurro, On the gradient inequality. Bull. Polish Acad. Sci. Math.
            47, 143–145 (1999)
[PdeM82]   J. Palis, W. de Melo, *Geometric Theory of Dynamical Systems* (Springer, Berlin, 1982)
[Sim83]   L. Simon, Asymptotics for a class of non-linear evolution equations, with applications
            to geometric problems. Ann. Math. **118**, 525–571 (1983)
[Tan75]   H. Tanabe, *Equations of Evolution* (Iwanami Shoten, Tokyo, 1975 (in Japanese)).
            (English translation: Pitman, London, 1979)
[Tan97]   H. Tanabe, *Functional Analytic Methods for Partial Differential Equations* (Dekker,
            New York, 1997)
[Tri78]   H. Triebel, *Interpolation Theory, Function Spaces, Differential Operators* (North-
            Holland, Amsterdam, 1978)
[Yag10]   A. Yagi, *Abstract Parabolic Evolution Equations and their Applications* (Springer,
            Berlin, 2010)
[Yag17]   A. Yagi, Real sectorial operators. Bull. South Ural State Univ. Ser. MMCS **10**, 97–112
            (2017)
[Yag]   A. Yagi, *Abstract Parabolic Evolution Equations and the Łojasiewicz–Simon Gradient
            Inequality, Applications*, vol. 2, (Springer, Berlin, to appear)
[YP14]   A. Yagi, M. Primicerio, A modified forest kinematic model. Vietnam J. Math. Anal. **12**,
            107–118 (2014)
[Yos80]   K. Yosida, *Functional Analysis,* 6th edn. (Springer, Berlin, 1980)
[Zei88]   E. Zeidler, *Nonlinear Functional Analysis and its Applications,* vol. I (Springer, Berlin,
            1988)

Symbol Index

© The Author(s), under exclusive license to Springer Nature Singapore Pte Ltd. 2021
A. Yagi, *Abstract Parabolic Evolution Equations and Łojasiewicz-Simon Inequality I*,
SpringerBriefs in Mathematics, https://doi.org/10.1007/978-981-16-1896-3

Subject Index

© The Author(s), under exclusive license to Springer Nature Singapore Pte Ltd. 2021
A. Yagi, *Abstract Parabolic Evolution Equations and Łojasiewicz-Simon Inequality I*,
SpringerBriefs in Mathematics, https://doi.org/10.1007/978-981-16-1896-3

Printed in the United States
by Baker & Taylor Publisher Services

Printed in the United States
by Baker & Taylor Publisher Services